T0091472

Seeing Four-Dimensional Space and Beyond

Using Knots!

SERIES ON KNOTS AND EVERYTHING

ISSN: 0219-9769

The Series on Knots and Everything: is a book series polarized around the theory of knots. Volume 1 in the series is Louis H Kauffman's Knots and Physics.

One purpose of this series is to continue the exploration of many of the themes indicated in Volume 1. These themes reach out beyond knot theory into physics, mathematics, logic, linguistics, philosophy, biology and practical experience. All of these outreaches have relations with knot theory when knot theory is regarded as a pivot or meeting place for apparently separate ideas. Knots act as such a pivotal place. We do not fully understand why this is so. The series represents stages in the exploration of this nexus.

Details of the titles in this series to date give a picture of the enterprise.

Published:

Vol. 75: *Scientific Legacy of Professor Zbigniew Oziewicz: Selected Papers from the International Conference "Applied Category Theory Graph-Operad-Logic"*
edited by H. M. C. García, José de Jesús Cruz Guzmán,
L. H. Kauffman & H. Makaruk

Vol. 74: *Seeing Four-Dimensional Space and Beyond: Using Knots!*
by E. Ogasa

Vol. 73: *One-Cocycles and Knot Invariants*
by T. Fiedler

Vol. 72: *Laws of Form: A Fiftieth Anniversary*
edited by L. H. Kauffman, F. Cummins, R. Dible, L. Conrad, G. Elisbury,
A. Crompton & F. Grote

Vol. 71: *The Geometry of the Universe*
by C. Rourke

Vol. 70: *Geometric Foundations of Design: Old and New*
by J. Kappraff

Vol. 69: *Mathematics of Harmony as a New Interdisciplinary Direction and "Golden" Paradigm of Modern Science*
Volume 3: The "Golden" Paradigm of Modern Science: Prerequisite for the "Golden" Revolution in Mathematics, Computer Science, and Theoretical Natural Sciences
by A. Stakhov

More information on this series can also be found at http://www.worldscientific.com/series/skae

K&E Series on Knots and Everything — Vol. 74

Seeing Four-Dimensional Space and Beyond

Using Knots!

Eiji Ogasa
Meiji Gakuin University, Japan

World Scientific

NEW JERSEY • LONDON • SINGAPORE • BEIJING • SHANGHAI • HONG KONG • TAIPEI • CHENNAI • TOKYO

Published by

World Scientific Publishing Co. Pte. Ltd.
5 Toh Tuck Link, Singapore 596224
USA office: 27 Warren Street, Suite 401-402, Hackensack, NJ 07601
UK office: 57 Shelton Street, Covent Garden, London WC2H 9HE

Library of Congress Control Number: 2023008261

British Library Cataloguing-in-Publication Data
A catalogue record for this book is available from the British Library.

Series on Knots and Everything — Vol. 74
SEEING FOUR-DIMENSIONAL SPACE AND BEYOND
Using Knots!

ISBN 978-981-127-512-8 (hardcover)
ISBN 978-981-127-515-9 (ebook for institutions)
ISBN 978-981-127-516-6 (ebook for individuals)

For any available supplementary material, please visit
https://www.worldscientific.com/worldscibooks/10.1142/13379#t=suppl

Typeset by Stallion Press
Email: enquiries@stallionpress.com

About the Author

Eiji Ogasa
Computer Science, Meijigakuin University, Yokohama, Kanagawa, 244-8539, Japan
ogasa@mail1.meijigkakuin.ac.jp
pqr100pqr100@yahoo.co.jp

Manuscripts of the author's papers can be obtained from his website, which can be found by typing the author's name, "Eiji Ogasa," into your search engine.

Contents

Acknowledgment

The author would like to thank the following persons:

Louis H. Kauffman
Liza Jacoby
Misaki Watanabe
Atsushi Mochizuki
Takayuki Nagasawa
Kokone Seya
Nanao Seya
Yukako Iwanaga
Hiroyuki Yokoi
Colin Adams

The names are written in no particular order.

Introduction

As you know, it is possible to imagine four-dimensional space and beyond. Mathematicians research these higher-dimensional spaces seriously. Moreover, using mathematics, people can actually visualize four-dimensional space and beyond. In this book, we introduce the tools that help us see higher-dimensional space.

According to string theorists, the physical elementary particles are almost certainly strings, and in fact, the world we live in is a 10- or 11-dimensional space or, a 10- or 11-dimensional manifold. Many readers interested in string theory may therefore be fond of higher-dimensional spaces and will find this guide to higher dimensions of interest.

In this book, we exhibit higher-dimensional space. We introduce a few kinds of manipulations of objects in higher-dimensional space.

For example, consider the object shown in Figure 0.1. Surely, you see a cube. However, isn't it simply a picture of three rhombi? We are using two-dimensional objects to construct what appears to be three-dimensional. Similarly, we can use figures drawn in the plane to help visualize four- and higher-dimensional objects.

Our way of thinking is a generalization of combinatorial knot theory. Readers do not need to know technical knot theory in Chapters 1–6; intuition for four-dimensional space and beyond will suffice. Undergraduate-level mathematics (linear algebra, for example) will be helpful.

In order to read the content in and after Chapter 7, the readers need to have knowledge from introductory textbooks on knot theory,

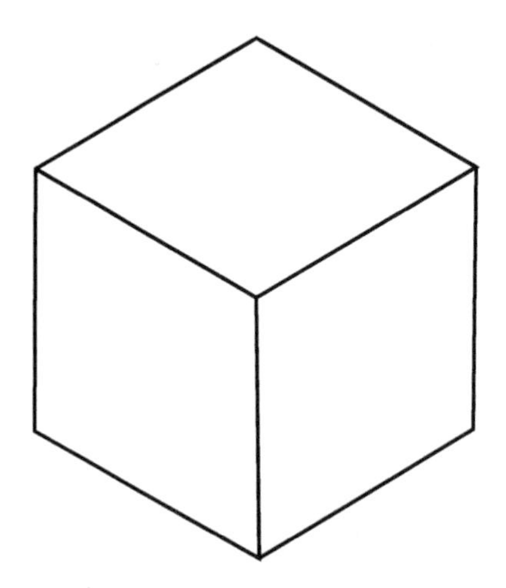

Figure 0.1 Three rhombi or a cubic?

e.g., Kauffman, *On Knots* [39] and Kauffman, *Knots and Physics* [40].

We also write about related important, recent topics: projections of knots, knot products, slice knots and links, Floer homology, and Khovanov–Lipshitz–Sarkar stable homotopy type.

Slice knots exist in 3-space, but are much related to 4-space. The slice problem is connected with many exciting topics: Khovanov homology, Khovanv–Lipshitz–Sarkar stable homotopy type, gauge theory, Floer homology, etc.

Khovanov–Lipshitz–Sarkar stable homotopy type is an exciting topic of recent development. It is defined for links in three-dimensional space and those in thickened surfaces, but it is a high-dimensional CW complex — a kind of higher-dimensional figure, in general. It is connected with string theory, topological quantum field theory, quantum algebra, and classical algebraic topology (classifying space, Steenrod square, etc.).

In this book, we additionally state relevant open questions. For example: Can the Jones polynomial be defined for links in all 3-manifolds?

Remark: In mathematical literature, n-dimensional space means $\mathbb{R}^n = \{(x_1, \ldots, x_n) \,|\, x_i$ is a real number$\}$ (see Section 1.1 in

Chapter 1, Section 2.1 in Chapter 2 and Section 4.3 in Chapter 4). However, "n-dimensional space" may often refer to the specific space $\mathbb{R}^n$, or to n-dimensional manifolds, or n-dimensional CW complexes. Note that the n-space $\mathbb{R}^n$ is both an n-dimensional manifold and an n-dimensional CW complex. Note that any n-dimensional manifold is an n-dimensional CW complex. See [74] for background reading on manifolds and CW complexes.

Chapter 1

Local Moves on Knots: For Beginners

There are numerous existing results on local moves on knots. We review some of them here.

In Chapters 1–6, we avoid using a lot of technical terms in the hope that the reader will understand the outline of ideas.

1.1 Knots

We begin by explaining knots.

Take *three-dimensional space* $\mathbb{R}^3$, as shown in Figure 1.1.

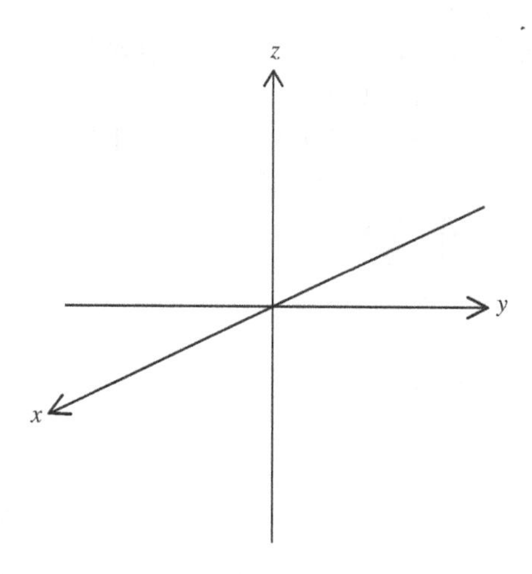

Figure 1.1 $\mathbb{R}^3$.

A single copy of a circle S^1 *embedded* in $\mathbb{R}^3$ is called a *knot.* See Figure 1.2. When we say that S^1 is *embedded* in $\mathbb{R}^3$, we mean that S^1 is included in $\mathbb{R}^3$ and S^1 does not touch itself. If a set of m copies of S^1 is embedded in $\mathbb{R}^3$, it is called an *m-component link.*

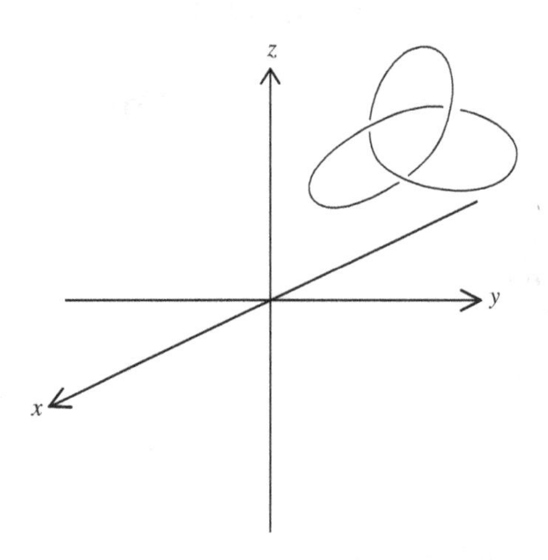

Figure 1.2 A knot in $\mathbb{R}^3$.

When we draw a knot, we often omit the x-, y-, and z-axes. We put a picture of a knot in $\mathbb{R}^2$ as in Figure 1.3. You must understand that a broken line indicates where one part of the curve undercrosses the other part. We call this kind of figure of knots a *knot diagram* or *diagram.*

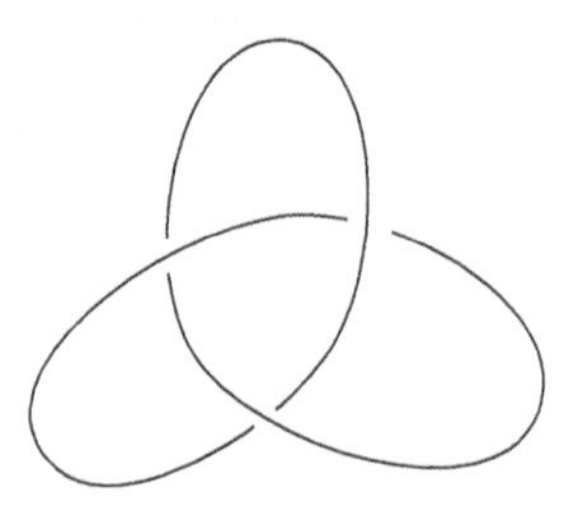

Figure 1.3 A knot diagram.

Consider the two knots shown in Figure 1.4. Do the two knots seem different?

Yes, in fact, they are.

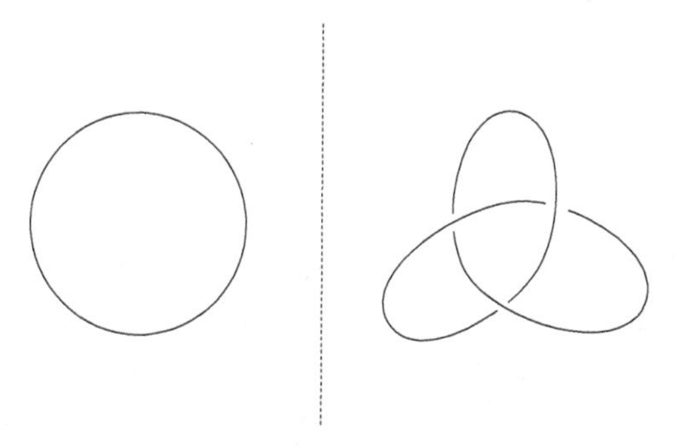

Figure 1.4 Two knots.

Let K and J be knots in $\mathbb{R}^3$. We say that K and J are the same, or (*smooth*) *isotopy*, if it is possible to manipulate K into J without touching or crossing itself: When we manipulate the knot, every part of it moves "smoothly."

We also say that J is obtained from K by a (*smooth*) *isotopy*.

Two knots in Figure 1.5 are isotopic. Note the crossing points.

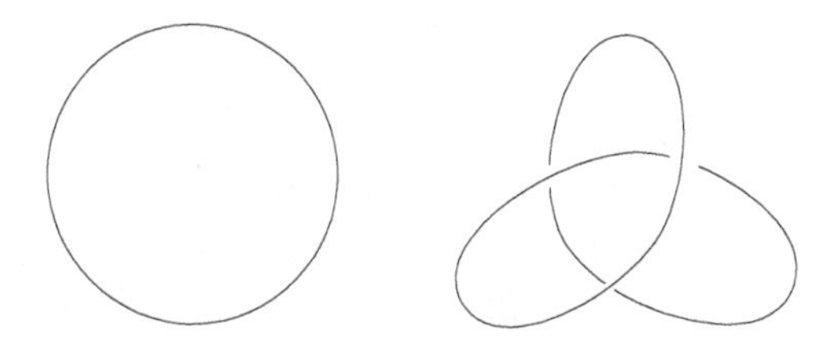

Figure 1.5 K is obtained from K' by an isotopy.

The *two-dimensional ball*, *2-ball*, *two-dimensional disc*, *2-disc*, or *disc*, is drawn in Figure 1.6. Let it be denoted by D^2 or B^2.

If a knot K bounds an embedded 2-ball B^2 in $\mathbb{R}^3$, we say that K is the *trivial knot* or the *unknot*. Note that if a knot J is istopic to the unknot, J is also the unknot.

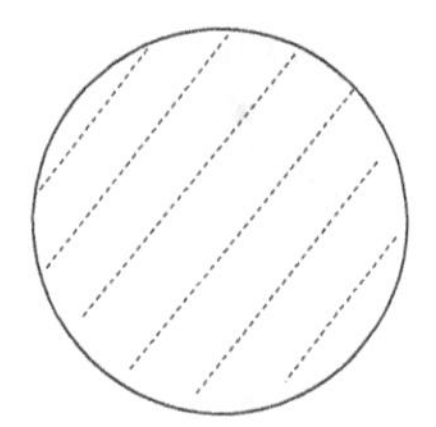

Figure 1.6 The disc.

Similarly, for an m-component link L, if each component bounds an embedded 2-ball B^2 in $\mathbb{R}^3$, and each B^2 does not touch any other, we say that L is the *trivial m-component link*. Note that if L' is isotopic to the trivial m-component link, L' is also the trivial m-component link.

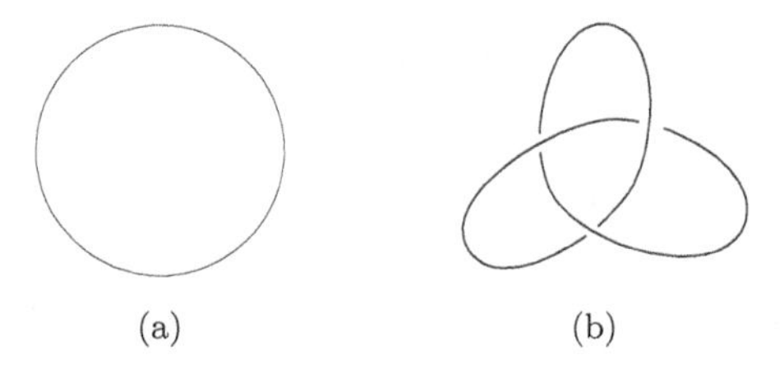

Figure 1.7 (a) The trivial knot and (b) the trefoil knot.

It is known that there are countably infinitely many different non-trivial knots. We describe in this book how to prove this fact.

The right knot in Figure 1.4 and the right knot in Figure 1.7 are the same non-trivial knot. It is called the *trefoil knot*.

1.2 Crossing Change

The *three-dimensional ball* or *3-ball* B^3 is drawn in Figure 1.8.

Any non-trivial knot can be changed into the unknot by a series of *crossing changes*, as shown in Figure 1.9. The proof is easy. Try.

If two knots K and K' in $\mathbb{R}^3$ differ only in the 3-ball B^3 as shown in Figure 1.9, then we say that K (respectively, K') is obtained from K' (respectively, K) by one crossing change. In Figure 1.10,

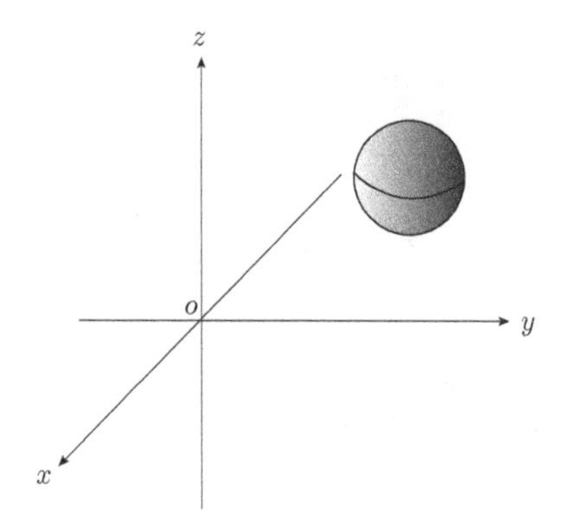

Figure 1.8 B^3.

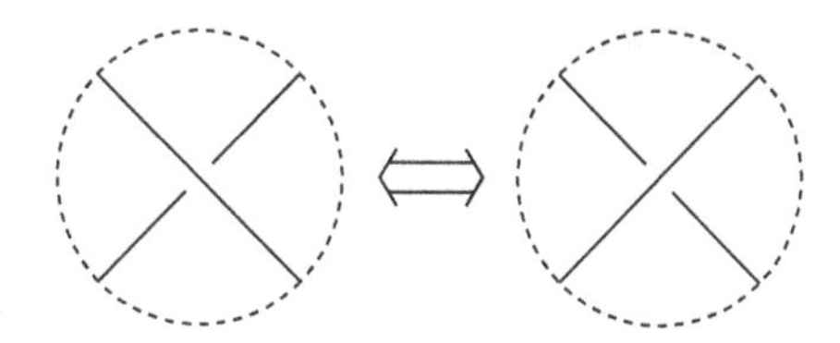

Figure 1.9 The crossing change of a knot.

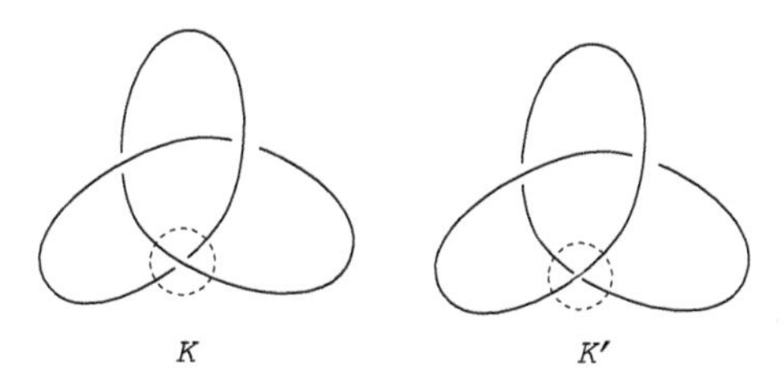

Figure 1.10 K is obtained from K' by one crossing change.

we obtain the trivial knot K' from the trefoil knot K by one crossing change.

Let J be a knot in $\mathbb{R}^3$. The minimum amount of crossing changes it takes to get the unknot is called the *unknotting number* of J.

Note: We may move a knot by an isotopy after and before we use an unknotting operation. See the sequence of Figures 1.11–1.14.

Figure 1.11 Moving a knot by an isotopy after and before the unknotting operation. This figure continues in Figure 1.12.

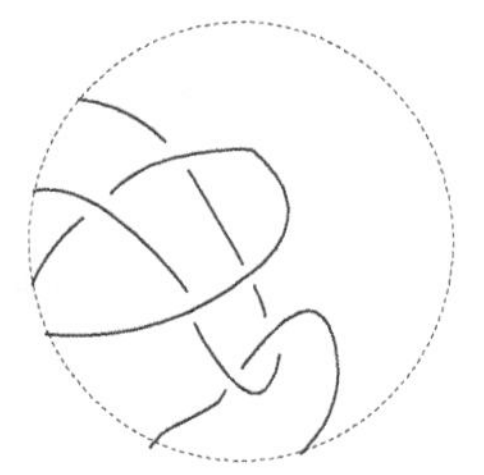

Figure 1.12 This figure continues in Figure 1.13.

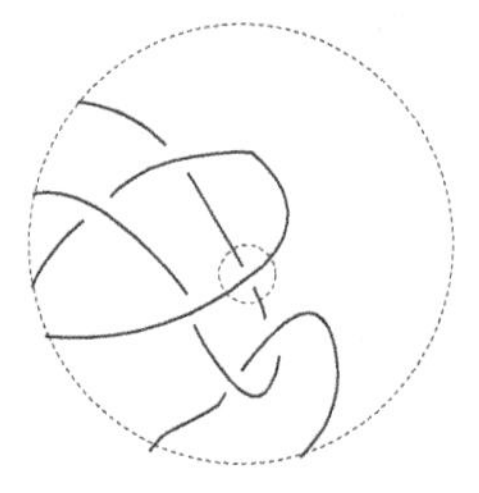

Figure 1.13 This figure continues in Figure 1.14.

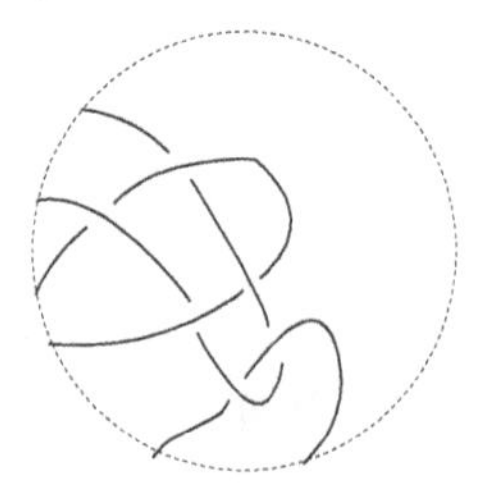

Figure 1.14 This figure is a continuation of Figure 1.13.

We have the following open question.

Question 1.1. Let K be an arbitrary knot in $\mathbb{R}^3$. Calculate the *unknotting number of* K.

For infinitely many kinds of knots, we know the answer to this question. However, there are also infinitely many knots for which we do not know the answer.

It is known that any m-component link is changed into the trivial m-component link by a finite sequence of crossing changes. The proof of this is left as an exercise.

1.3 Pass-Move

We have a type of manipulation called a *pass-move* on links. See Figure 1.15 for an illustration of it.

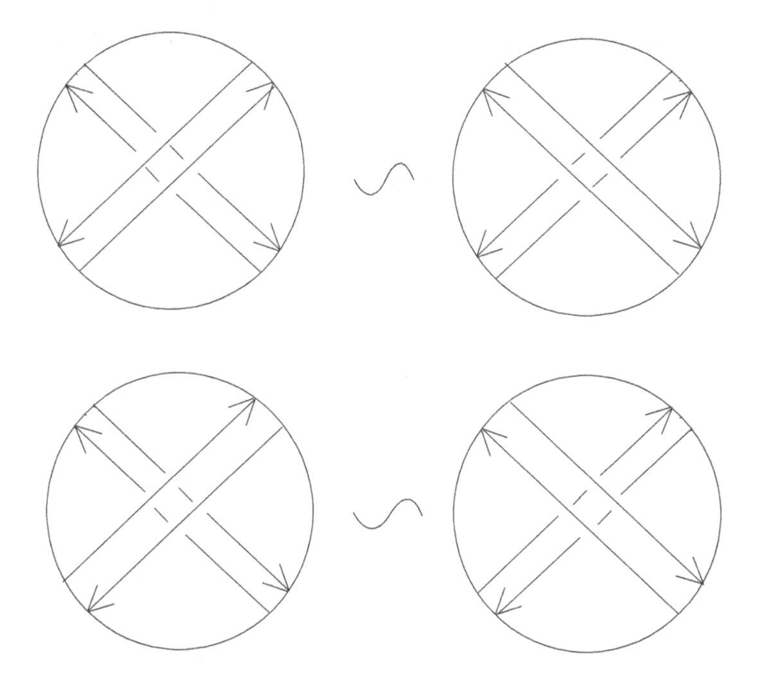

Figure 1.15 Pass-move.

For this type of manipulation, we consider not only knots but also links. We orient the link; the orientation is represented by the arrows in Figure 1.15. In Section 6.1 of Chapter 6, we discuss the orientation of knots and links for detail.

When we consider a link with more than one component, each of the four arcs in the 3-ball may belong to a different component.

We show an example of a pass-move in Figures 1.16–1.18.

Note that in Figure 1.16, we draw a one-component link and not a 2-component link. Moreover, it is the trivial knot.

Now you try: Carry out a pass-move on this 1-knot in of Figure 1.17.

You may suspect that the resulting knot in Figure 1.18 is a non-trivial knot. Yes, this is true!

It is a known fact that there are countably infinitely many non-trivial knots which are pass-move equivalent to the trivial knot. In fact, any knot is either pass-move equivalent to the trivial knot or to the trefoil knot, although they are not pass-move equivalent to one another.

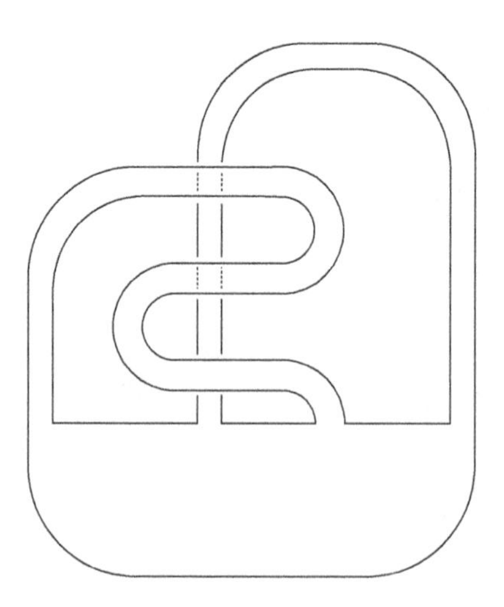

Figure 1.16 Note that this is the trivial one-component link or the trivial knot.

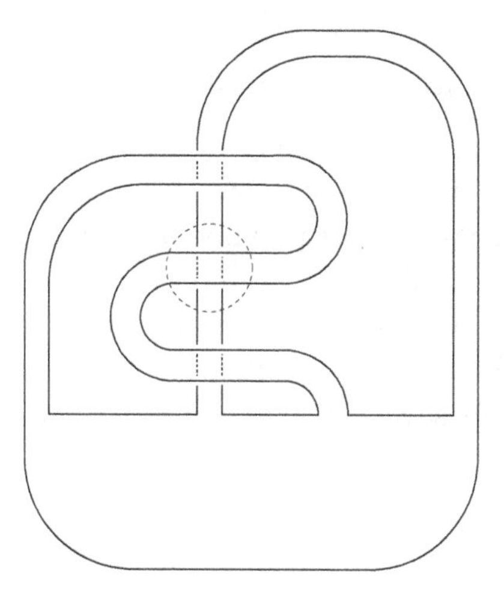

Figure 1.17 This knot is the same as that in Figure 1.16.

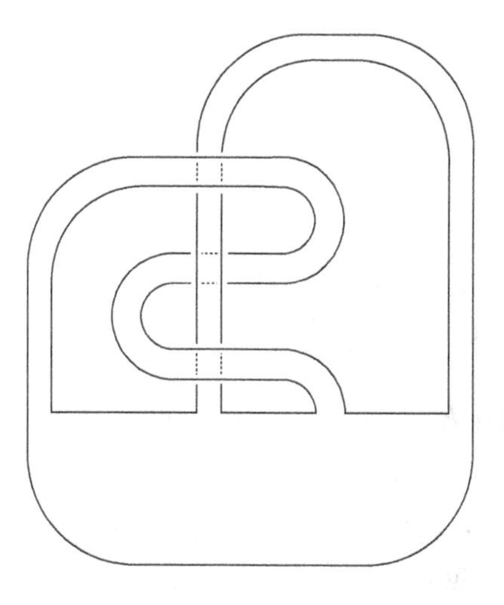

Figure 1.18 A pass-move make this knot from that in Figure 1.17.

For any knot K, we can define the *Arf invariant* — either 0 or 1. If the Arf invariant of K is 0 (respectively, 1), then K is pass-move equivalent to the trivial knot (respectively, the trefoil knot). See [39].

1.4 Well-Known Local Move Identities for the Alexander and Jones Polynomials

In knot theory, we study what knots exist and what properties those Dknots have. We have two well-known tools to investigate knots and links: the Alexander and Jones polynomials. In each case, we assign to a knot a polynomial. We discuss the properties of these polynomials here. In Chapter 6, we introduce the strict definitions.

Assume that three links only differ in a particular 3-ball, as shown in Figure 1.19. We say that these links are related by the *skein relation.*

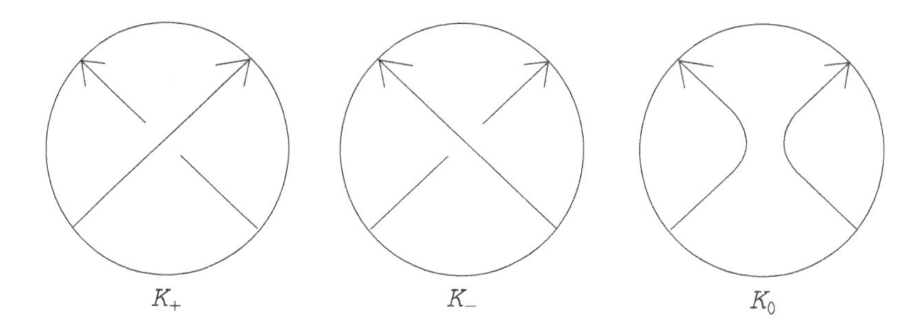

Figure 1.19 The skein relation.

To every knot, we assign an Alexander polynomial, whose formal definition we provide later on in Chapter 6: We use the skein relation to define them.

If K_+, K_-, and K_0 are related in the skein relation, we have

$$\text{Alex}(K_+) - \text{Alex}(K_-) = (t^{\frac{1}{2}} - t^{-\frac{1}{2}}) \cdot \text{Alex}(K_0).$$

In Figure 1.20, we show an example of three links related by the skein relation. In Figure 1.20, K_+ is the trefoil knot. The Alexander polynomial of the trefoil knot is $t - 1 + t^{-1}$. In Figure 1.20, K_- is the trivial knot. The Alexander polynomial of the trivial knot is 1. In Figure 1.20, K_0 is a 2-component link called the *Hopf link*, named after the great mathematician Heinz Hopf. Its Alexander polynomial is $t^{\frac{1}{2}} - t^{-\frac{1}{2}}$.

In the case of Figure 1.20, combining the Alexander polynomials of the trefoil and trivial knots for the left-hand side of the Alexander

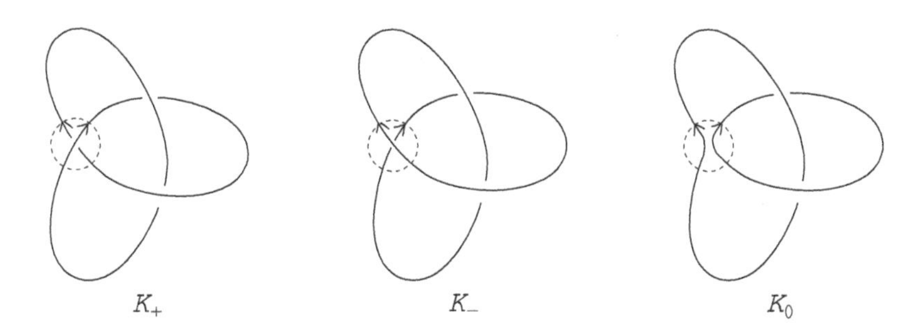

Figure 1.20 An example of the skein relation.

identity, we have $(t - 1 + t^{-1}) - 1$. On the other hand, for the right side of the identity, we have $(t^{\frac{1}{2}} - t^{-\frac{1}{2}})(t^{\frac{1}{2}} - t^{-\frac{1}{2}})$. We see that these two are equal.

We also assign to each knot the Jones polynomial. We will write the definition, later, in Chapter 6. If K_+, K_-, and K_0 are related in the skein relation, we have

$$t^{-1}V(K_+) - tV(K_-) = (t^{\frac{1}{2}} - t^{-\frac{1}{2}})V(K_0).$$

Return to Figure 1.20. Therefore, three links in Figure 1.20 satisfy this identity.

The Jones polynomial of K_+ is $t + t^3 - t^4$ and that of K_- is 1. For K_0, we have $-t^{\frac{1}{2}} - t^{\frac{5}{2}}$.

The left-hand side of the Jones identity is

$$t^{-1}(t + t^3 - t^4) - t = 1 + t^2 - t^3 - t.$$

The right-hand side of the identity is

$$(t^{\frac{1}{2}} - t^{\frac{-1}{2}})(-t^{\frac{1}{2}} - t^{\frac{5}{2}}) = -t + 1 - t^3 + t^2.$$

Again, we see that both sides are equal.

You may be curious: Are there any local move operations in higher-dimensional space? Furthermore, is there a relationship between knot polynomials and local move operations in the high-dimensional case? The answer to both of these questions is yes.

We go on to explain it in the following chapters.

Chapter 2

Four-Dimensional Space $\mathbb{R}^4$

2.1 Four-Dimensional Space $\mathbb{R}^4$

We provide an overview of four-dimensional Euclidean space $\mathbb{R}^4$. In this section, the reader will find useful key concepts from an undergraduate linear algebra course. Though, even without the knowledge of linear algebra, readers can easily understand the following if they have an interest in sci-fi novels, movies, animes, and comics.

The set $\{(x, y, z)\}$ of all ordered triples of real numbers x, y, and z is called *three-dimensional space*, or *3-space*, denoted by $\mathbb{R}^3$. Formally, we write

$$\mathbb{R}^3 = \{(x, y, z)|x, y, \text{ and } z \text{ are arbitrary real numbers}\}.$$

Similarly, the set $\{(x, y, z, w)\}$ of all ordered 4-tuples of arbitrary real numbers x, y, z, and t is called *four-dimensional space*, or *4-space*, denoted by $\mathbb{R}^4$. Formally, we have

$$\mathbb{R}^4 = \{(x, y, z, t)|x, y, z, \text{ and } t \text{ are arbitrary real numbers}\}.$$

We draw $\mathbb{R}^4$ as shown in Figure 2.1.

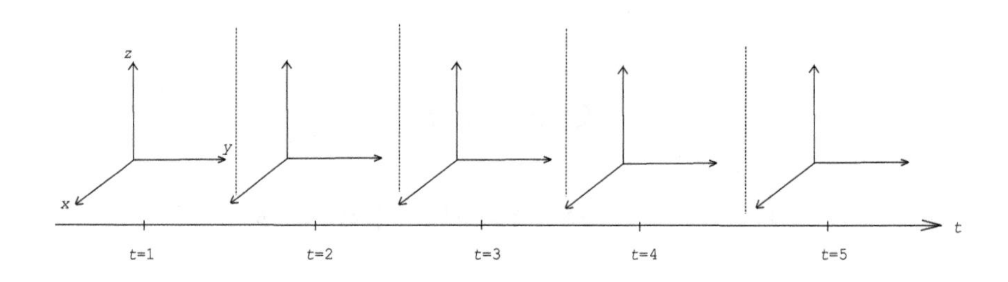

Figure 2.1 $\mathbb{R}^4$.

In Figure 2.2, we show the set $A = \{(x, y, z, t)|x = y = z = 1, 2 \leqq t \leqq 4\}$ in $\mathbb{R}^4$. What is A?

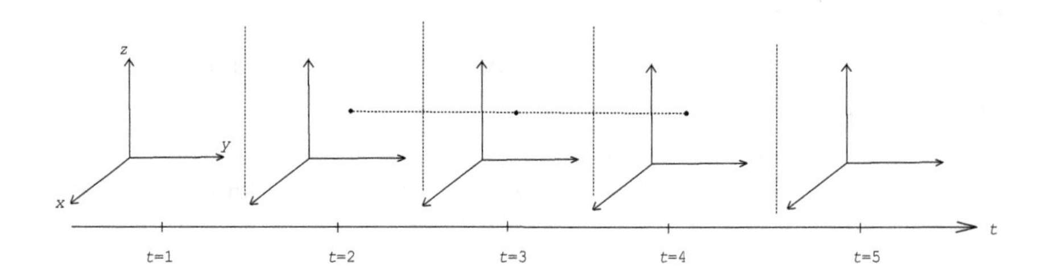

Figure 2.2 What is A?

A is a segment. We can move A around in $\mathbb{R}^4$: The result is shown in Figure 2.3.

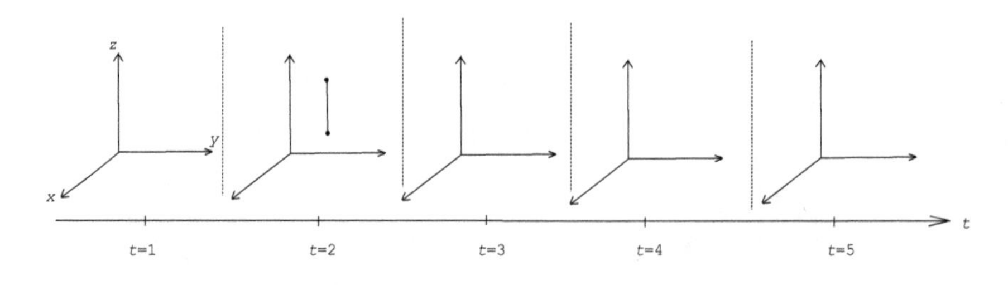

Figure 2.3 A is ...

As for moving segments in $\mathbb{R}^4$, like moving the segment from Figure 2.2 to what's shown in Figure 2.3, we have a one-dimension-higher analog of the following process.

Draw a segment in $\mathbb{R}^3$, as shown in Figure 2.4.

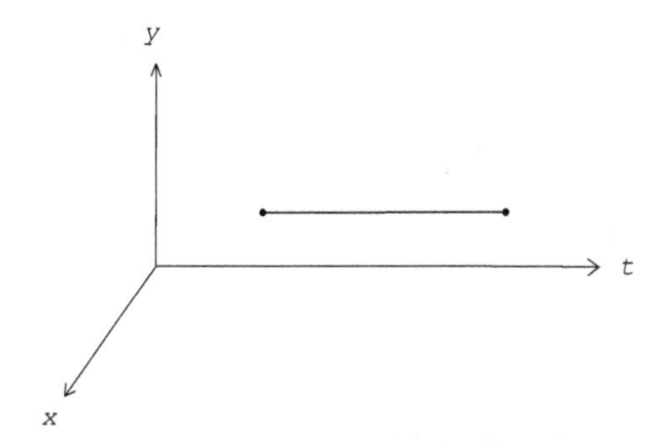

Figure 2.4 A segment in $\mathbb{R}^3$.

Consider the planes perpendicular to the t-axis intersecting the segment, shown in Figure 2.5, and the sections of our segment contained in these intersecting planes.

Sections $\mathbb{R}^2$ at $t = 1, 2, 3, 4, 5$ in $\mathbb{R}^3$ in Figure 2.5 correspond to sections $\mathbb{R}^3$ at $t = 1, 2, 3, 4, 5$ in $\mathbb{R}^4$ in Figure 2.2.

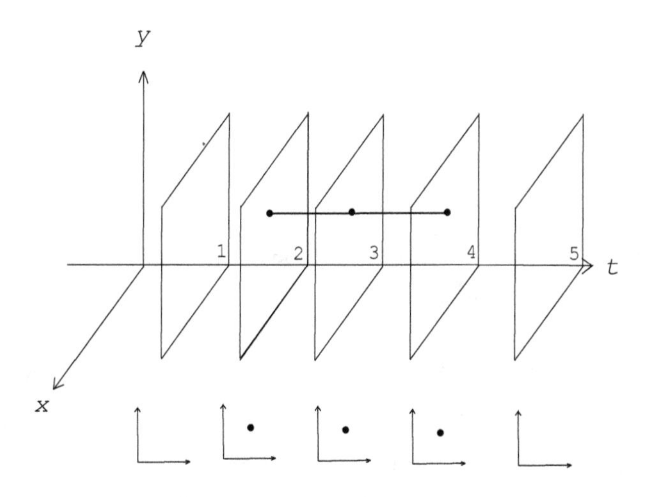

Figure 2.5 Sections of a segment in $\mathbb{R}^3$.

We move the segment from Figure 2.6,

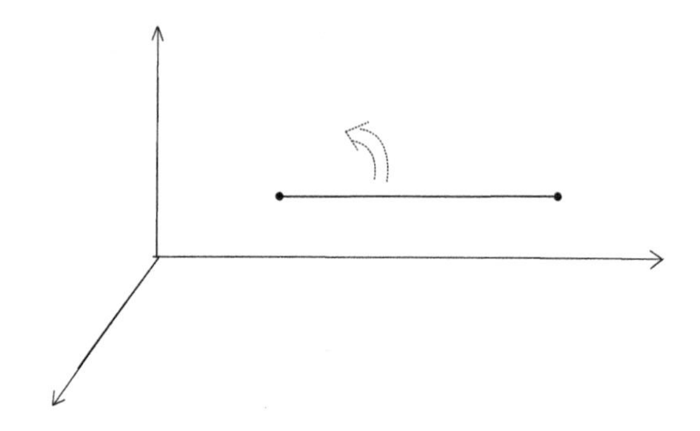

Figure 2.6 Moving a segment in $\mathbb{R}^3$.

and place it as it appears in Figure 2.7.

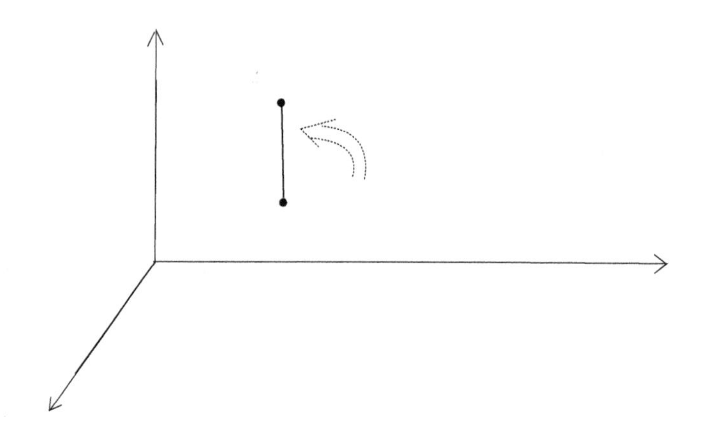

Figure 2.7 The result of the moving.

Now that we have a better understanding of 4-space, we can go on to discuss knots and links.

2.2 A Non-Trivial Link in 3-Space Which Can Be Untied in 4-Space

The link in the right image of Figure 2.8 is the trivial 2-component link. The one on the left is non-trivial, and the Hopf link.

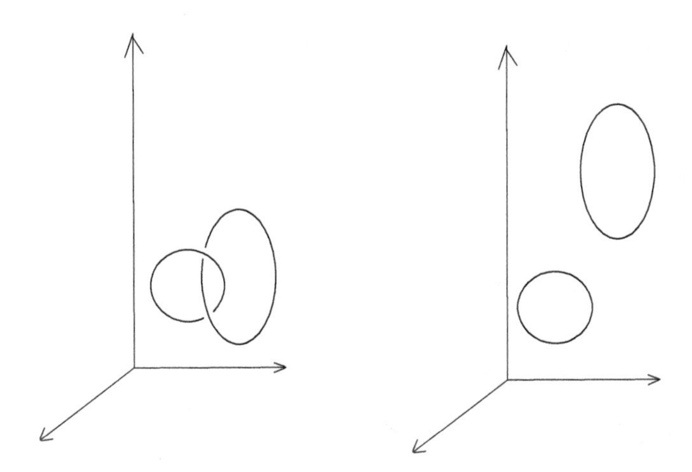

Figure 2.8 The left figure is the Hopf link. The right figure is the trivial 2-component link.

In $\mathbb{R}^3$, we cannot unlink the Hopf link — that is, we cannot use an isotopy to get from the Hopf link to the trivial link in $\mathbb{R}^3$. However, we can do it in $\mathbb{R}^4$. We explain it as follows.

More strictly, we explicate the following method for unlinking the Hopf link in 4-space. We suppose that both the Hopf link and the trivial 2-component link are embedded in

$$\begin{aligned}\mathbb{R}^3 &= \{(t,x,y,z)|\, t=1, x, y, \text{ and } z \text{ are arbitrary real numbers}\}\\ &= \{(1,x,y,z)|\, x, y, \text{ and } z \text{ are arbitrary real numbers}\},\end{aligned}$$

a subspace of

$$\mathbb{R}^4 = \{(t,x,y,z)|\, t, x, y, \text{ and } z \text{ are arbitrary real numbers}\}.$$

We claim that we can move the Hopf link in $\mathbb{R}^4$ by an isotopy to obtain the trivial 2-component link.

Take the Hopf link in $\mathbb{R}^3$ at $t = 1$.

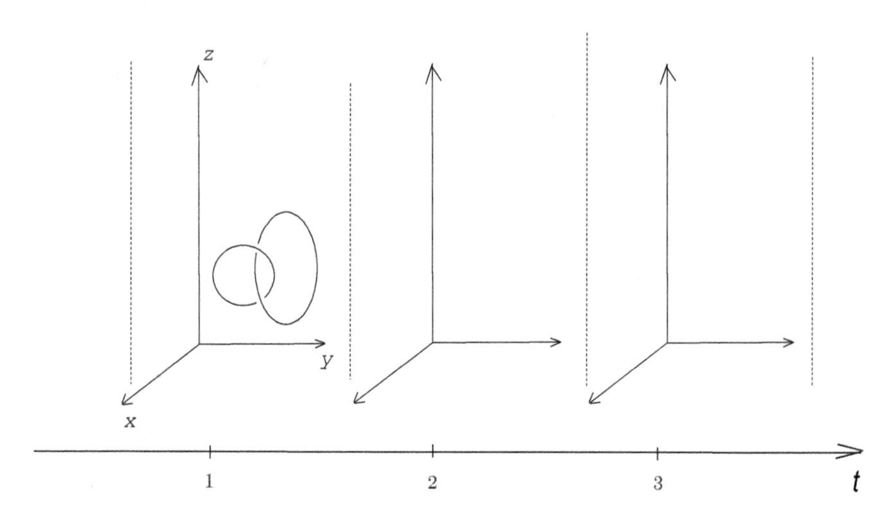

Choose one component — say, the right-hand one — and move it from $t = 1$ to $t = 3$. All the while, the left-hand component of the Hopf link at $t = 1$ stays put.

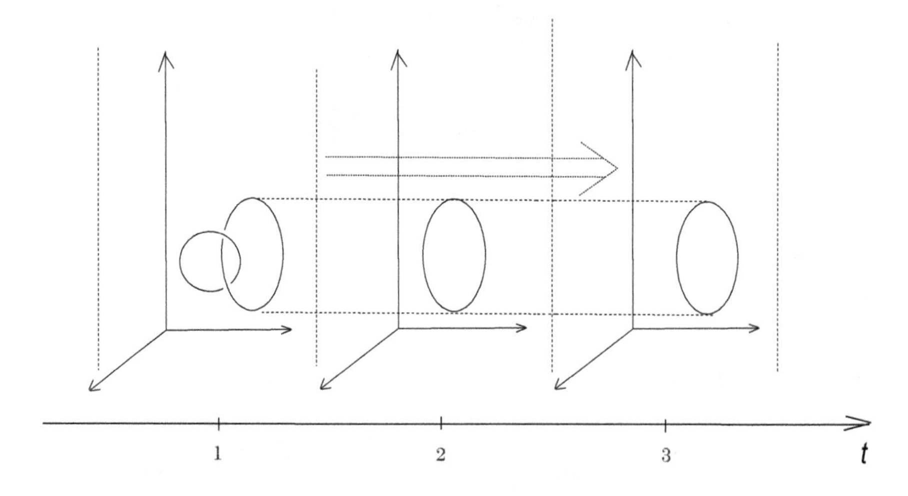

The result is as drawn in the following figure.

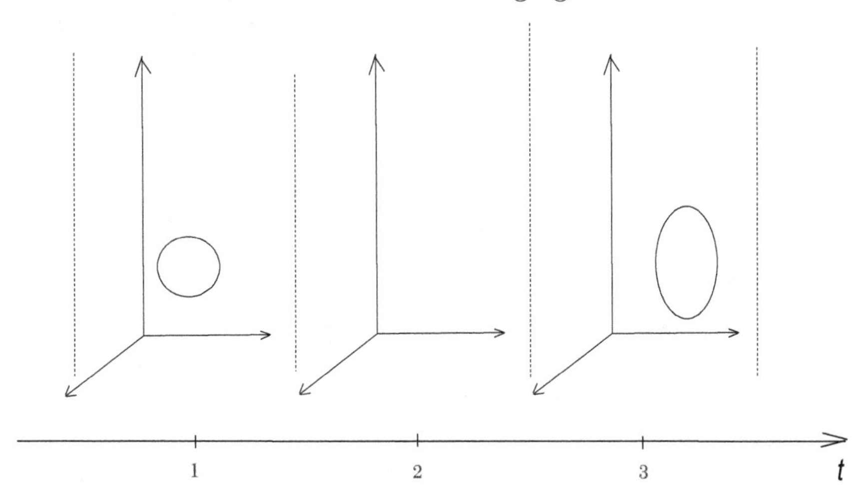

We now have two components at different t values — one in $\mathbb{R}^3$ at $t = 1$, and one in $\mathbb{R}^3$ at $t = 3$. Take the component at $t = 3$ and move it like so, keeping the other component in place.

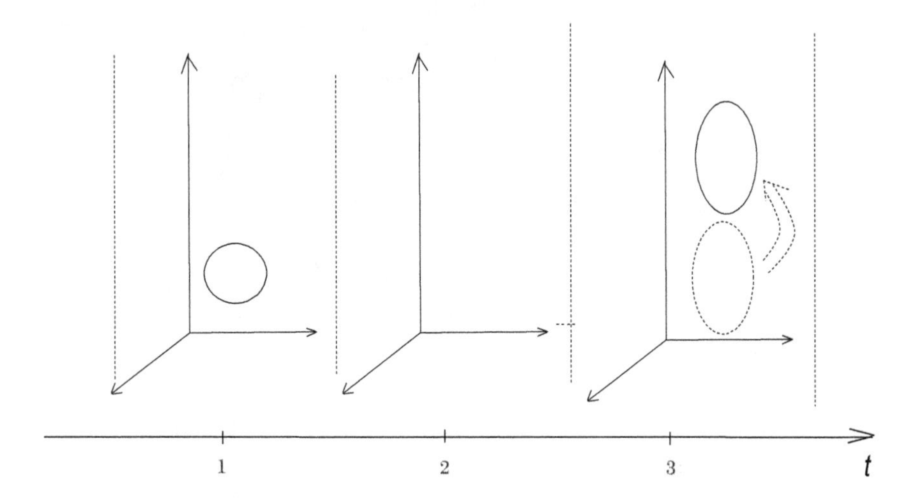

The result is as drawn in the following figure.

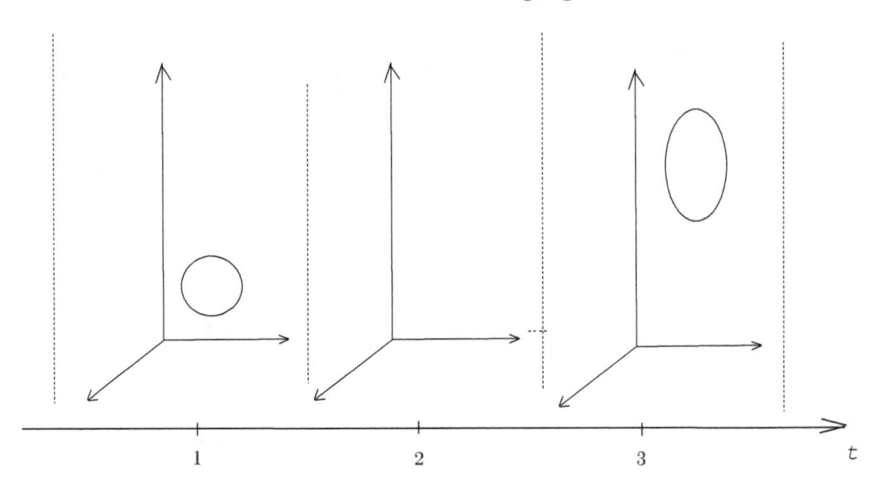

Now we again move the $t = 3$ component, but this time back to $t = 1$. The $t = 1$ component still remains in place.

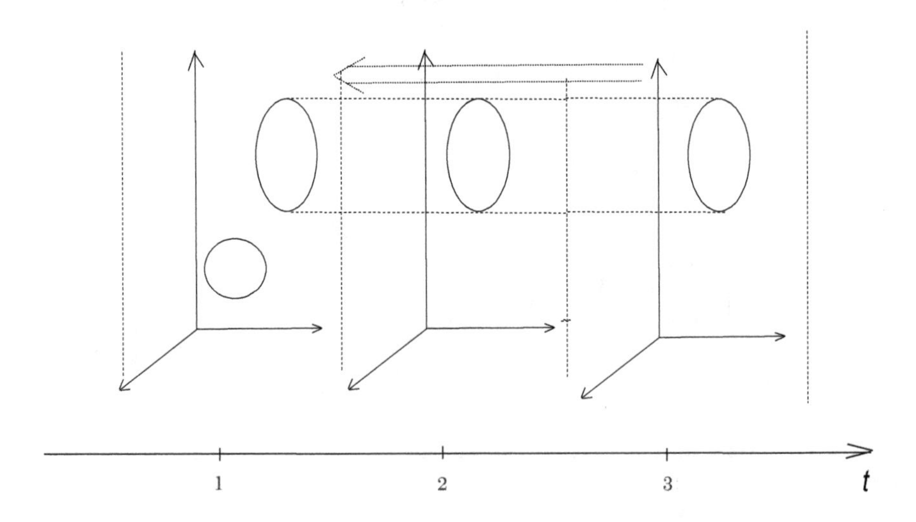

The result is shown in the following figure; we have the trivial 2-component link in $\mathbb{R}^3$ at $t = 1$.

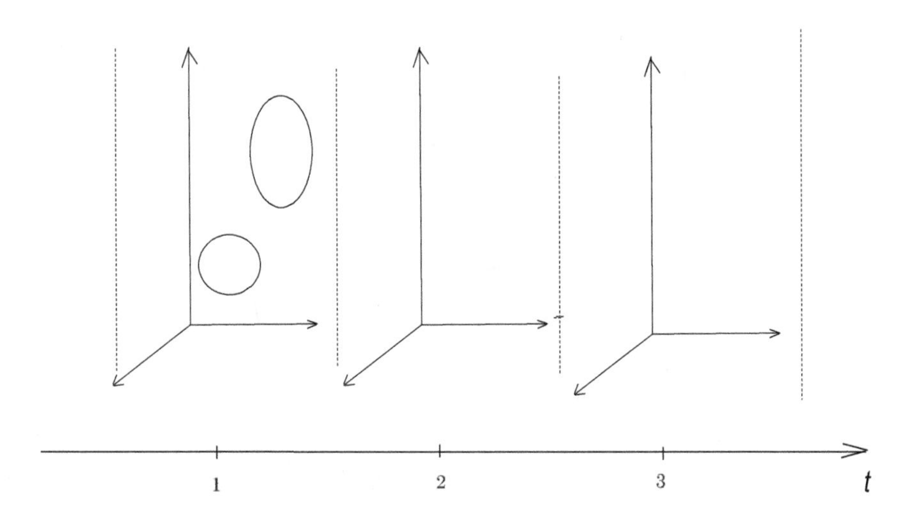

We succeeded! We unlinked the Hopf link.

The unlinking operation we just outlined is merely a one-dimension-higher analog of the following process, which is perhaps more visualizable.

Draw a circle in the plane $\mathbb{R}^2$ as in Figure 2.9. Although we smoothly transform a circle without touching or crossing itself, we also call the resulting figure a circle. Take a point A in the circle and a point B outside the circle. Can you move the point A over to the point B without touching or crossing over the circle? You may think this is impossible, and you would be right.

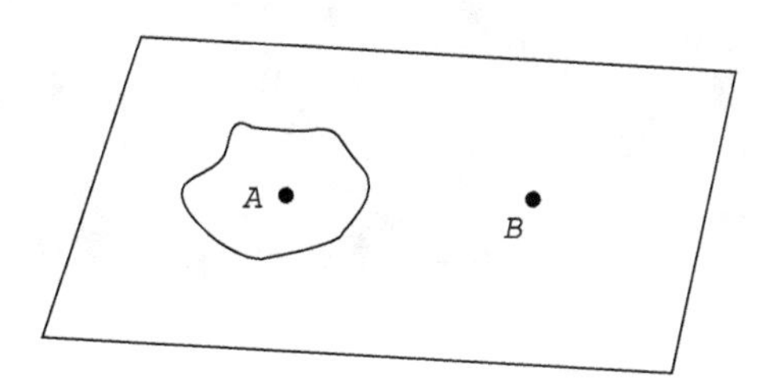

Figure 2.9 A circle and two points in the plane.

However, what if we can go outside of the plane? Then we can easily move point A to point B. Figure 2.10 shows a path between

A and B that goes above the plane containing the two points. While we cannot use such a path in $\mathbb{R}^2$, this is perfectly legal in $\mathbb{R}^3$.

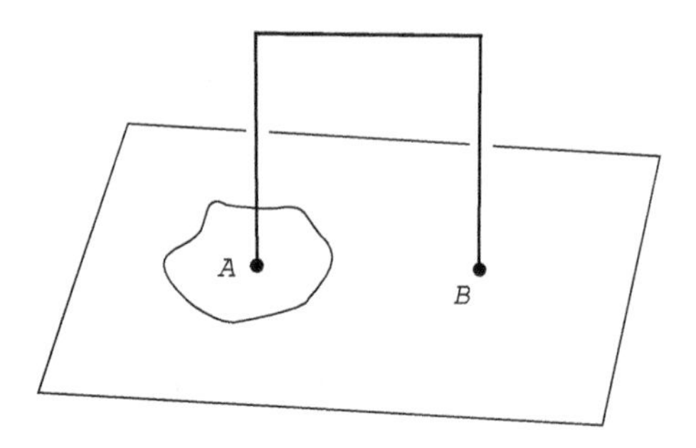

Figure 2.10 Using the outside of the plane.

Tea break. Genius sisters discovered a great fact about the Hopf link (Figure 2.11).

Figure 2.11 We can make the Hopf link!

You can find a movie of theirs by typing "Eiji Ogasa" in your search engine or YouTube.

2.3 Untying One-Dimensional Knots in 4-Space

All one-dimensional knots in 3-space can be unknotted in 4-space. For example, consider the trefoil knot in Figure 2.12. We certainly cannot unknot the trefoil — that is, use isotopy to get from the trefoil to the unknot — in $\mathbb{R}^3$.

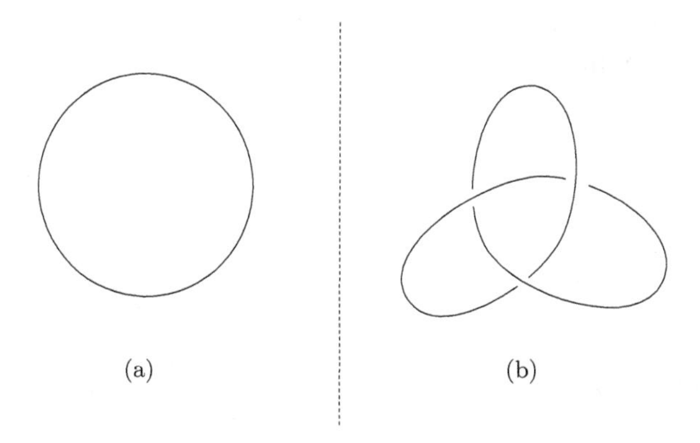

Figure 2.12 (a) The trivial knot and (b) the trefoil knot 12.

However, we can do it in $\mathbb{R}^4$. We explain it as follows.

More strictly, we outline a method similar to the one for links in the previous section. Assume that both the trefoil knot and the trivial knot are embedded in

$$\begin{aligned}\mathbb{R}^3 &= \{(t, x, y, z) | \, t = 1, x, y, \text{ and } z \text{ are arbitrary real numbers}\} \\ &= \{(1, x, y, z) | \, x, y, \text{ and } z \text{ are arbitrary real numbers}\},\end{aligned}$$

a subspace of

$$\mathbb{R}^4 = \{(t, x, y, z) | \, t, x, y, \text{ and } z \text{ are arbitrary real numbers}\}.$$

We claim that just as we can unlink the Hopf link in $\mathbb{R}^4$, we can manipulate the trefoil knot in $\mathbb{R}^4$ by isotopy to achieve the trivial knot.

First, we discuss an operation which is made possible in the plane by looking at it as a subspace of 3-space. The method for untying the trefoil knot is an analog of this operation, but in one-dimension-higher.

Consider the left-hand side of Figure 2.13, which shows a circle in the plane around a point. Can we manipulate the circle in the plane to get the image on the right-hand side of Figure 2.13 without ever touching the point?

You may be thinking this cannot work. And you're right — this sort of manipulation is not possible in $\mathbb{R}^2$.

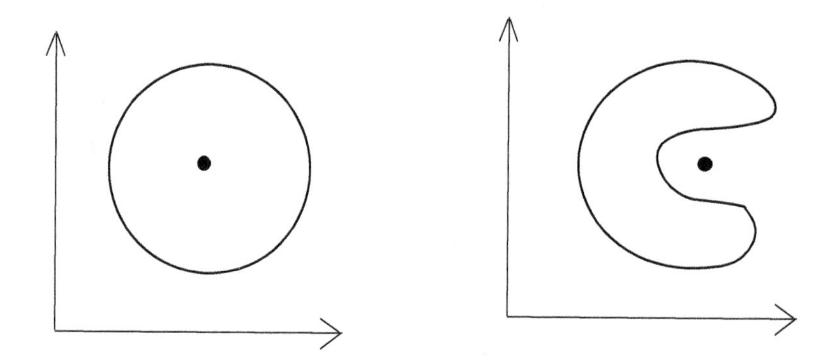

Figure 2.13 We cannot do this operation in the plane.

However, we can look at this problem in $\mathbb{R}^3$ as shown in Figure 2.14. In this case, we can remove the point from the circle. First move the point outside the plane. Move the circle as you please, then push the point back to the original plane. We now have exactly what we wanted.

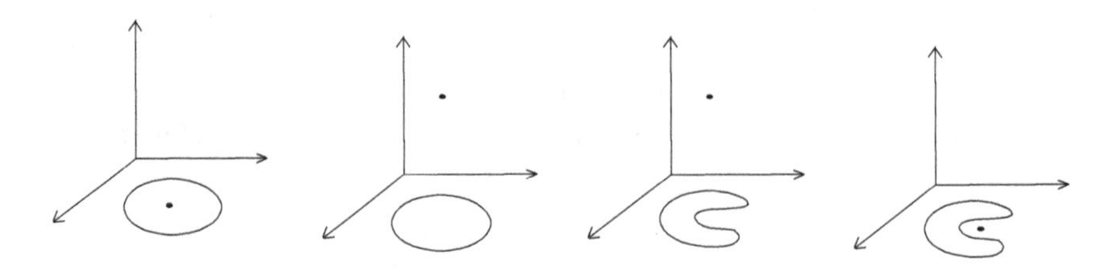

Figure 2.14 Using the outside of the plane.

It is also useful to consider the following figures. We draw sections of the objects in Figures 2.15 and 2.16 obtained from planes perpendicular to the t-axis. For untying knots, we consider an analog of this notion in one-dimension-higher.

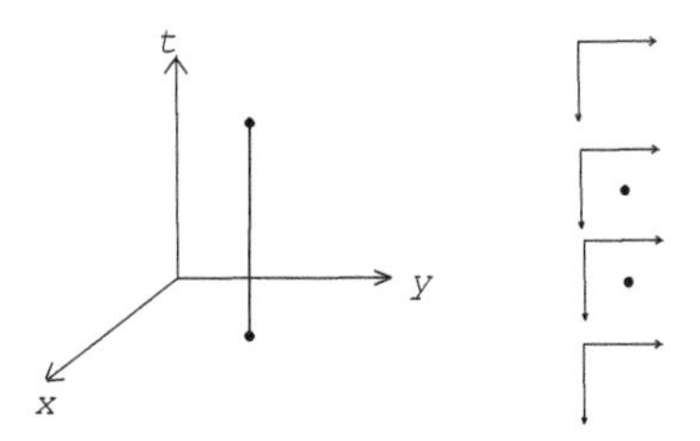

Figure 2.15 Sections of the segment in $\mathbb{R}^3$.

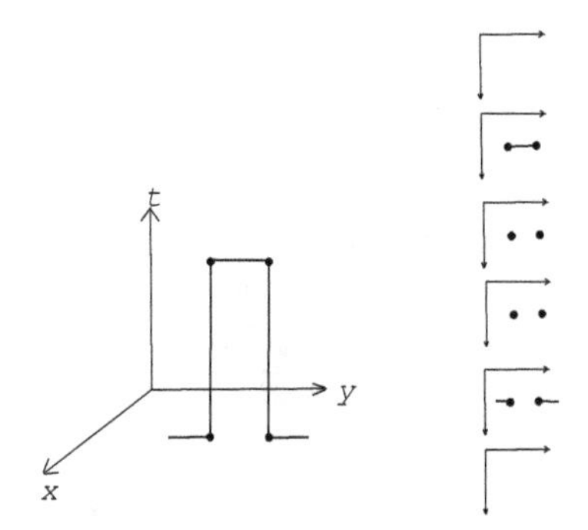

Figure 2.16 Sections of an object in $\mathbb{R}^3$.

We now have the tools to untie the trefoil knot in $\mathbb{R}^4$. Embed the trefoil in $\mathbb{R}^3$ as a subspace of $\mathbb{R}^4$, at $t = 2$.

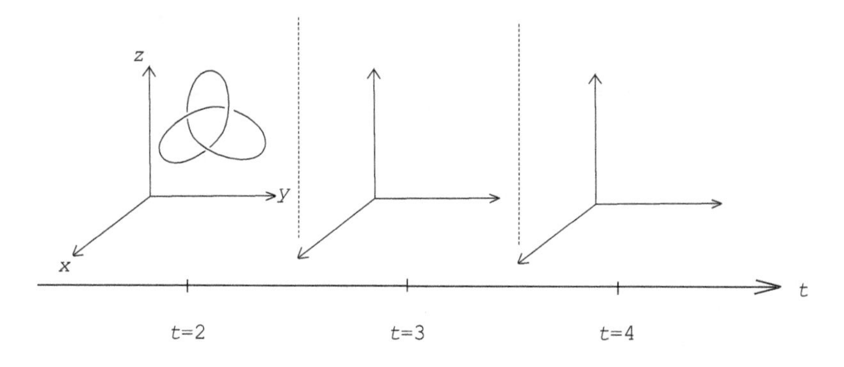

Move a curved segment in the trefoil knot into $t > 2$ as drawn in the following figure. Note that in $2 < t < 3$, the trefoil knot has two parts, which are equivalent to two open intervals.

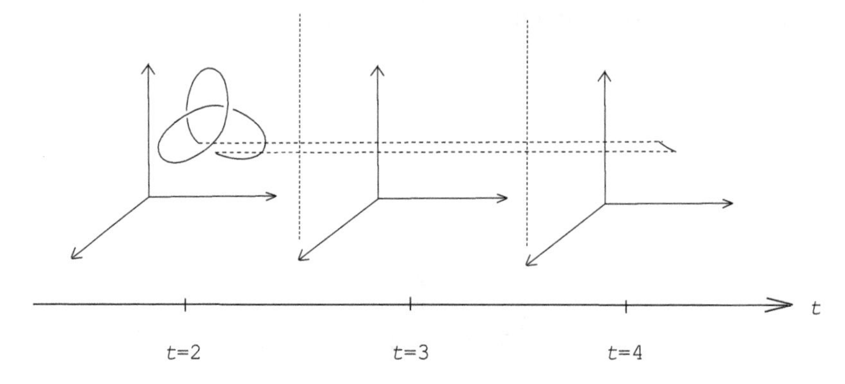

We can move one part from $t = 2$ as shown in the following figure.

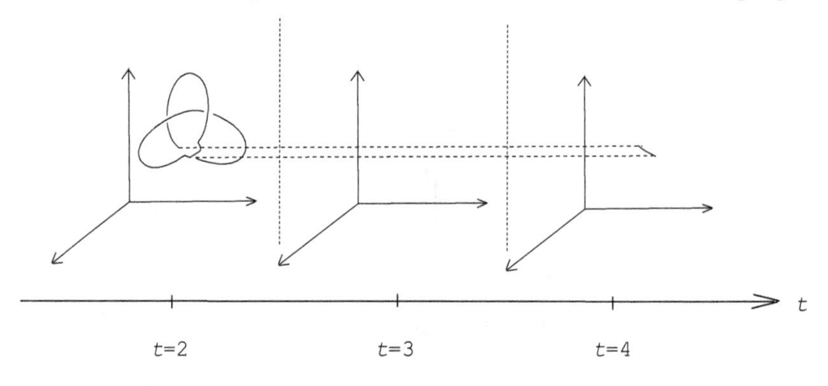

Then push back the part in $t > 2$ as shown in the following figure.

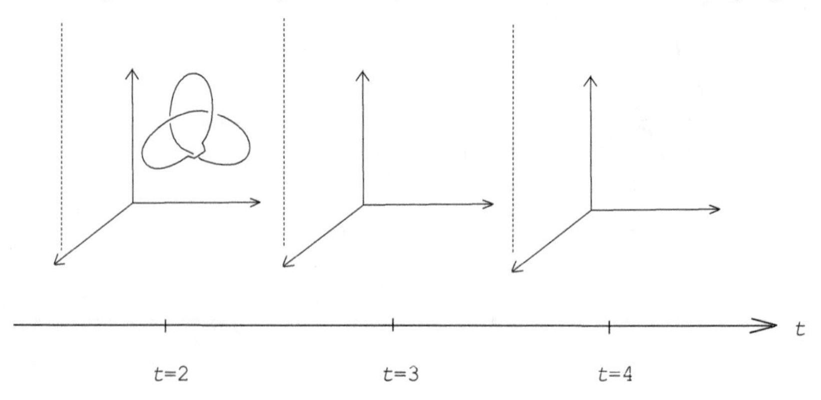

The result is the knot in $\mathbb{R}^3$ at $t = 2$ shown in the preceding figure. Note that this knot is in fact the trivial knot.

We can repeat this method for any one-dimensional knot embedded in $\mathbb{R}^3$. Hence, any such knot can be unknotted in $\mathbb{R}^4$. We can use this method on every component of a link in $\mathbb{R}^3$, and consequently, combining what we've learned here with the method outlined in Section 2.2, any link in $\mathbb{R}^3$ can be made into the trivial link in $\mathbb{R}^4$.

Note that in fact, we can make all links into the trivial link by using only the method of this section. The proof is easy. Try.

2.4 Knotted Objects in 4-Space and Beyond

In Sections 2.2 and 2.3, we elucidated how all knots and links embedded in $\mathbb{R}^3$ can be made into the trivial knot or link in $\mathbb{R}^4$. But you may be wondering: Are there knots (or knotted objects) embedded in 4-space? What about higher dimensions?

Yes, such objects do exist! We discuss them in the following chapters.

Chapter 3

Local Moves in Higher-Dimensional Space: For Beginners

3.1 Manipulation of Objects in Three-Dimensional Space

We describe the case of knots in $\mathbb{R}^3$ before we introduce the higher-dimensional case.

As an example, we change the trivial 2-component link in Figure 3.1

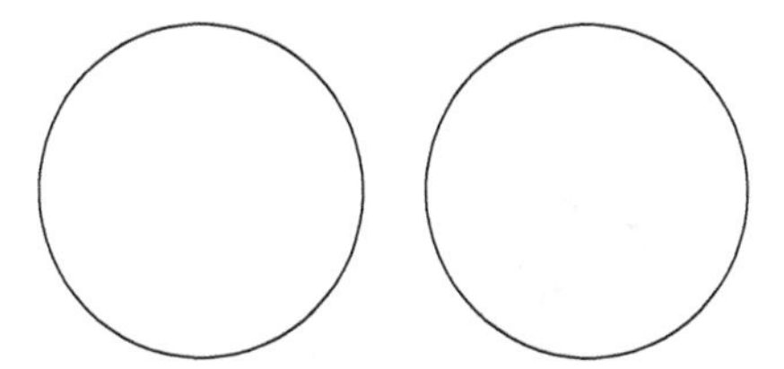

Figure 3.1 The trivial 2-component link.

into the Hopf link in Figure 3.2

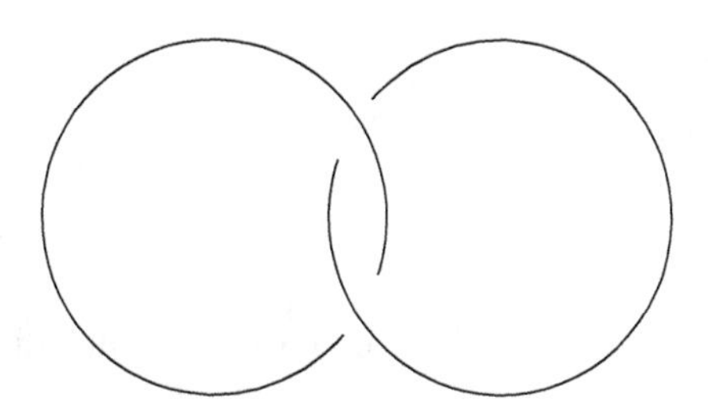

Figure 3.2 The Hopf link.

by one crossing change. This is shown in Figure 3.3.

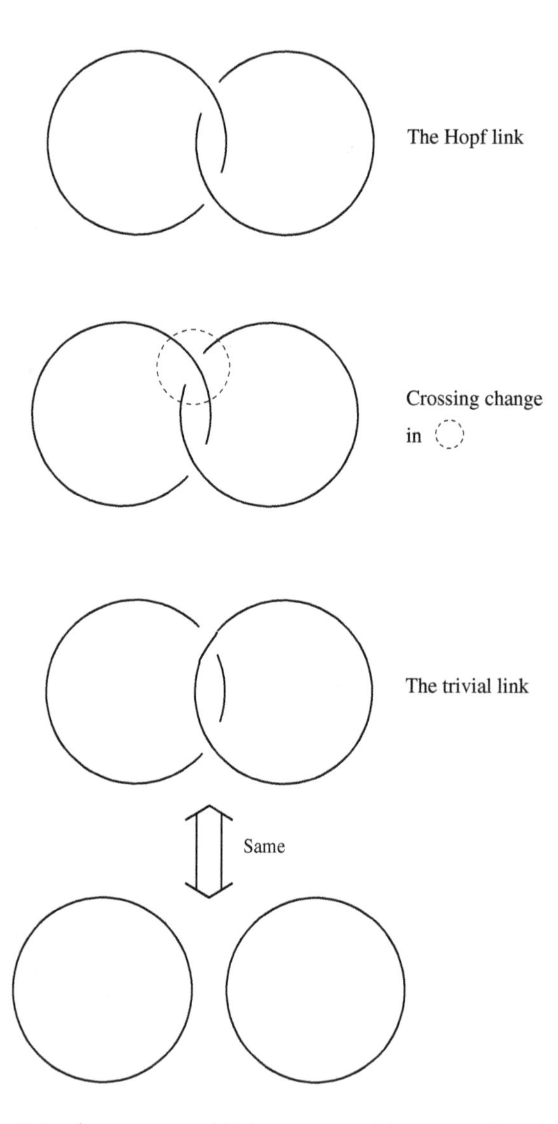

Figure 3.3 A sequence of links generated by a crossing change.

This crossing change operation can be generalized into higher-dimensional space.

The two links shown in Figure 3.3 have different properties, as we explain in the following.

Take a disc D^2 embedded in $\mathbb{R}^3$. The boundary of D^2 is one of the two components of the trivial 2-component link (respectively, the Hopf ink).

Can we choose our disc so that it does not intersect the other component of the link?

For the trivial link, absolutely. For the Hopf link, this is not possible.

3.2 Ribbon-Moves on a Disjoint Union of a Circle and a Sphere in $\mathbb{R}^4$

For a circle S^1 and a sphere S^2, we denote their *disjoint union* as $S^1 \amalg S^2$. Take $S^1 \amalg S^2$ in $\mathbb{R}^3$ as shown in Figure 3.4. Note that S^2 does not touch S^1 (hence "disjoint") and that neither object touches itself.

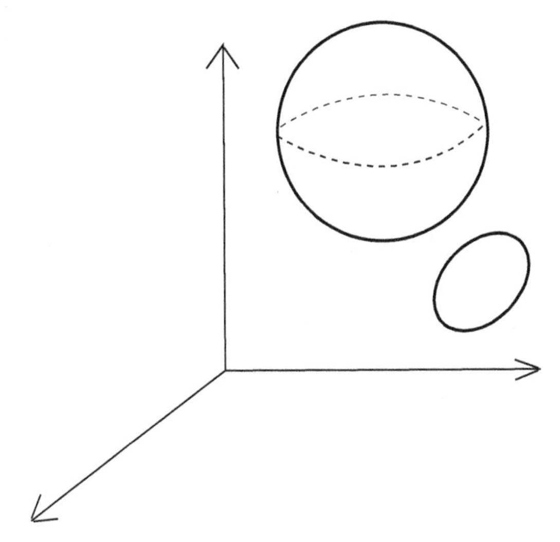

Figure 3.4 Disjoint union of S^1 and S^2.

Next, we construct two versions of the disjoint union $S^1 \amalg S^2$ in $\mathbb{R}^4$ — call them L_0 and L_1 — such that neither S^1 nor S^2 touches itself.

L_0: S^1 bounds an embedded two-dimensional disc D^2 in $\mathbb{R}^4$.
S^2 bounds an embedded three-dimensional ball B^3 in $\mathbb{R}^4$.
B^3 does not touch D^2.

If you're thinking that the existence of L_0 is pretty believable, you're right! It does exist. Recall that we can visualize $\mathbb{R}^4$ as the $xyzt$-space, with $\mathbb{R}^3$ — the xyz-space — as a subspace. Consider in $\mathbb{R}^3$ as a subspace of $\mathbb{R}^4$ the disjoint union of "D^2 with S^1" and "B^3 with D^2." For reference, see Figure 3.5.

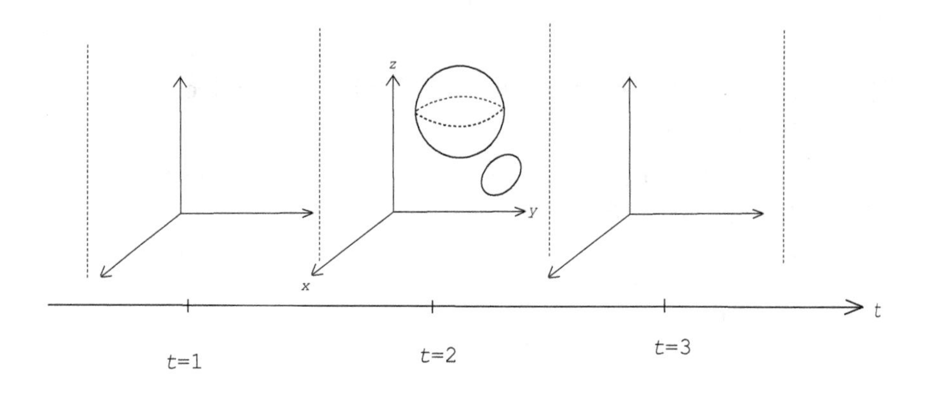

Figure 3.5 L_0.

We now construct L_1 as described in the following:

L_1: S^1 bounds an embedded two-dimensional disc D^2 in $\mathbb{R}^4$, but we allow D^2 to intersect S^2.

S^2 bounds an embedded three-dimensional ball B^3 in $\mathbb{R}^4$, but we allow B^3 to intersect S^1.

No matter how we put D^2, it intersects S^2. But, we can choose D^2 so that the intersection is a single point.

In fact, such an L_1 does exist! Can you imagine L_1? Can you construct it?

Before attempting to construct L_1, recall the following fact from Section 3.2: the trivial 2-component link can be embedded in both $\mathbb{R}^2$ and $\mathbb{R}^3$, but the Hopf link can only be embedded in $\mathbb{R}^3$, not $\mathbb{R}^2$; we can carry out one crossing change on the trivial link in $\mathbb{R}^3$ to obtain the Hopf link. See Figure 3.6 for reference.

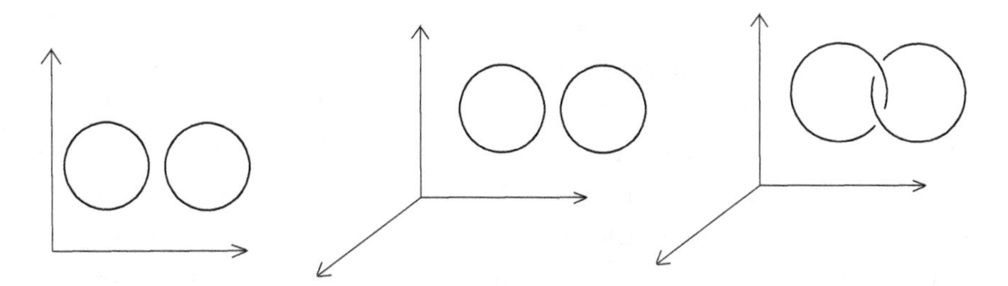

Figure 3.6 The left and middle images are the trivial 2-component link. The right one is the Hopf link.

Before drawing L_1 in $\mathbb{R}^4$, let's go through a construction of L_0 in $\mathbb{R}^4$ different from that shown in Figure 3.5.

Put S^1 and S^2 in $\mathbb{R}^4$ as shown in Figure 3.7.

S^1 exists in $1 \leqq t \leqq 3$:

$S^1 \cap \mathbb{R}^3$ at $t = 1$ is a segment.
$S^1 \cap \mathbb{R}^3$ at each t of $1 < t < 3$ is two points.
$S^1 \cap \mathbb{R}^3$ at $t = 3$ is a segment.

The circle S^1 in this case is drawn as a rectangle with four corners, we still regard it as a circle. The reason is that we can smoothly transform a rectangle without touching itself, into a circle. Note that in Figure 3.7, S^1 and S^2 in $\mathbb{R}^4$ fit the description of L_0.

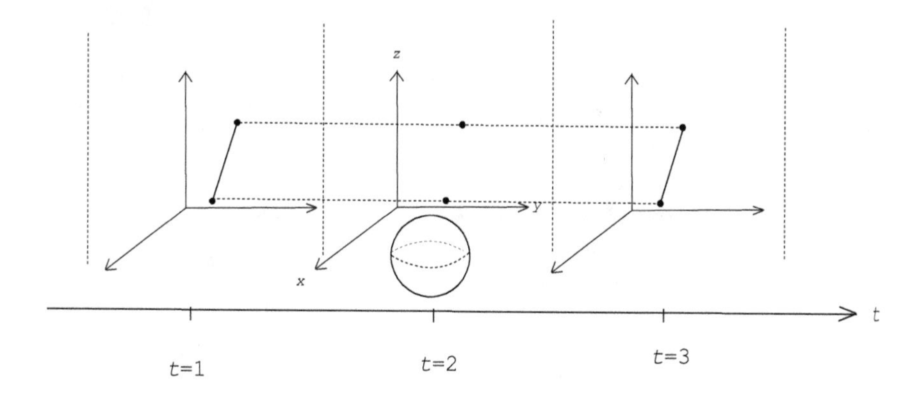

Figure 3.7 L_0.

Let's construct L_1. Put S^1 and S^2 in $\mathbb{R}^4$ as shown in Figure 3.8, placing S^1 in the same position as that shown in Figure 3.7. $S^1 \cap (\mathbb{R}^3$ at $t = 2)$ is two points, one of which exists in the inside of S^2 in $\mathbb{R}^3$ at $t = 2$, and the other, in the outside. Recall that in $\mathbb{R}^2$, S^1 bounds a disk, separating the inside of the circle from the outside. In $\mathbb{R}^3$, that is not the case — the inside of the circle is connected to the outside. Similarly, when we embed S^2 in $\mathbb{R}^4$, there is no separation between the part of space inside the sphere and the part of space that is outside. In Figure 3.8, S^1 and S^2 make L_1.

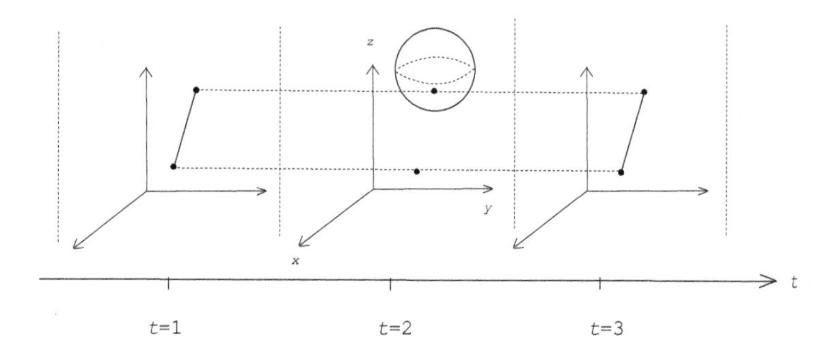

Figure 3.8 L_1.

It is known that L_0 and L_1 are different. (They are not isotopic.) The above L_0 and L_1 in $\mathbb{R}^4$ are drawn in Figure 3.9.

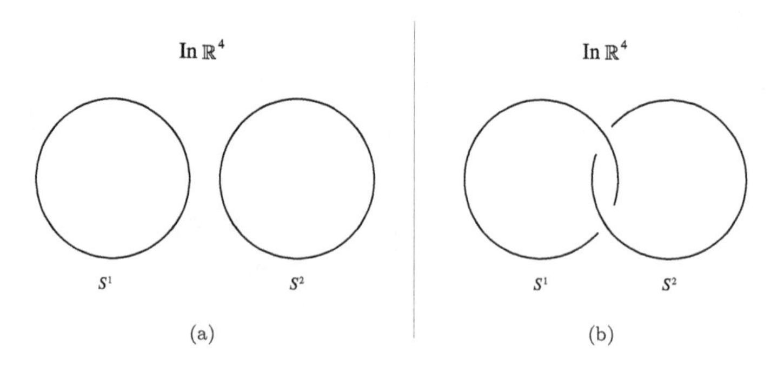

Figure 3.9 (a) L_0 and (b) L_1 in $\mathbb{R}^4$.

Note that we can change L_1 into L_0 by one local move, as shown in Figure 3.10, which is a one-dimension-higher analog of Figure 3.3.

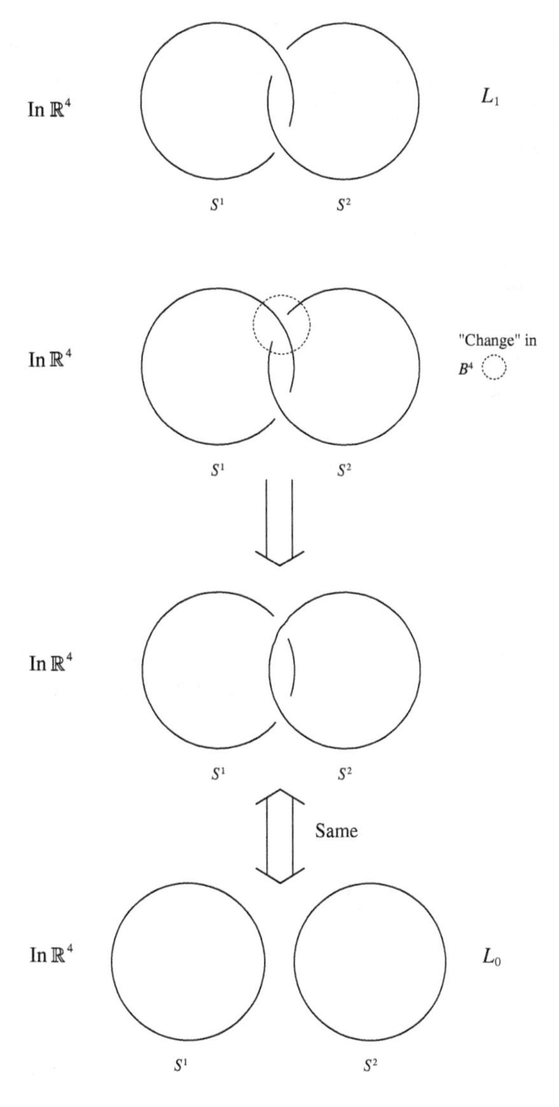

Figure 3.10 A sequence of $S^1 \amalg S^2$ in $\mathbb{R}^4$ generated by a ribbon-move.

We define the *four-dimensional ball* or *4-ball* as

$$B^4 = \{(x, y, z, t) \subset \mathbb{R}^4 | x^2 + y^2 + z^2 + t^2 \leqq 1\}.$$

In Figure 3.10, (dotted circle) is a 4-ball showing the local move which changes L_1 to L_0. This 4-ball is shown in Figure 3.11. Let B^1 denote a segment in our 4-ball; B^1 is sometimes called the 1-ball. Recall that B^2 is the 2-ball. The procedure shown is called the *ribbon-move*, and we say that L_1 (respectively, L_0) is changed into L_0 (respectively, L_1) by one ribbon-move.

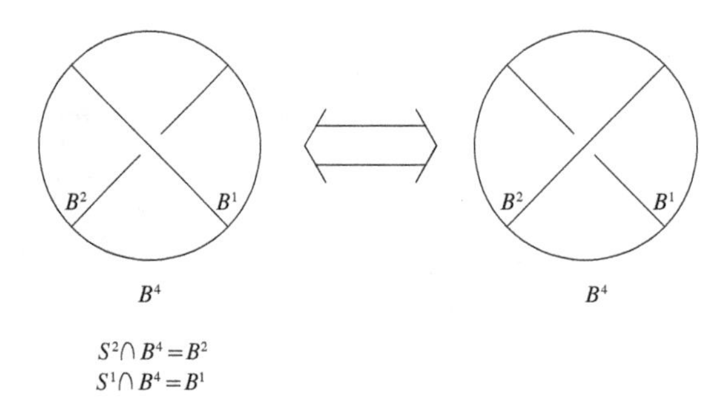

Figure 3.11 The ribbon-move.

In Figures 3.12 and 3.13, we show more explicitly the action that is occurring in the ribbon-move. We change the part in Figure 3.12 into that in Figure 3.13.

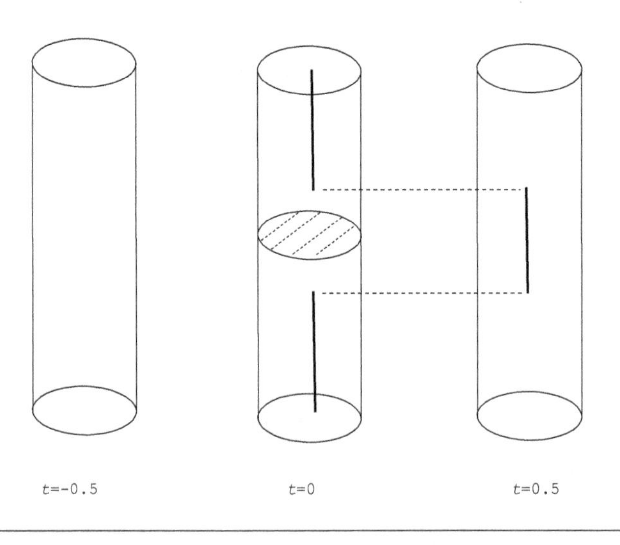

Figure 3.12 The ribbon-move. This figure continues in Figure 3.13.

Now, we embed B^4 in 4-space — $txyz$-space — as follows:

We regard $B^4 \cap (\mathbb{R}^3$ at $t = -1)$ as a stuffed or solid cylinder. Note that a solid cylinder is transformed smoothly without touching itself, into the 3-ball. A solid cylinder moves from $t = -1$ to $t = 1$; the trace is B^4.

We only draw sections $t = -0.5, 0, 0.5$. The 4-ball B^4 includes a disjoint union of the 2-ball B^2 and the segment B^1 as drawn in Figures 3.12 and 3.13. Note that by our construction, B^4 has a corner, but we can still regard it as B^4 since we can continuously smooth out the corner without touching it, to get what we typically imagine as B^4.

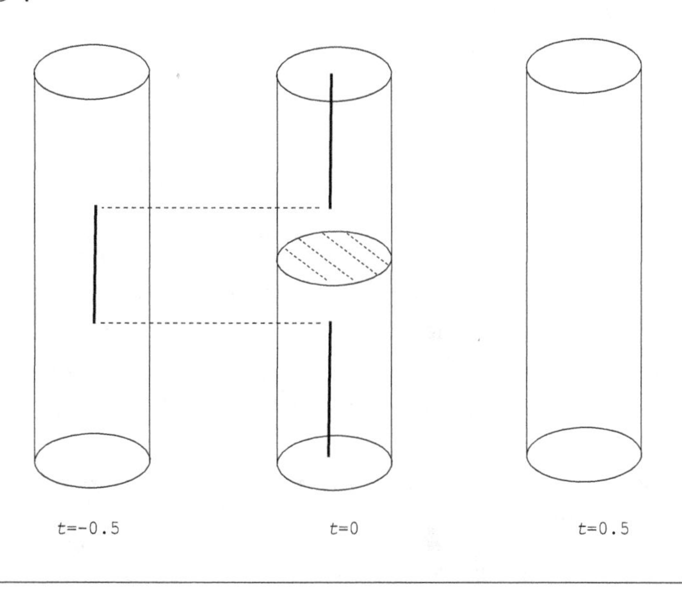

Figure 3.13 The ribbon-move. This figure is a continuation of Figure 3.12.

For the reader who prefers literal expressions, we can write this explicitly as follows:

$$B^4 = \{(t, x, y, z) | -1 \leqq t \leqq 1, 0 \leqq z \leqq 1, x^2 + y^2 \leqq 1\}.$$

B^2 is embedded in the $txyz$-space as follows:

$$\left\{(t, x, y, z) | t = 0, z = \frac{1}{2}, x^2 + y^2 \leqq 1\right\}.$$

B^1 in Figure 3.12 is

$$\left\{(t,x,y,z)|t=0, 0 \leqq z \leqq \frac{1}{3}, \frac{2}{3} \leqq z \leqq 1, x=y=0\right\}$$
$$\cup \left\{(t,x,y,z)|0 \leqq t \leqq 0.5, z=\frac{1}{3}, z=\frac{2}{3}, x=y=0\right\}$$
$$\cup \left\{(t,x,y,z)|t=0.5, \frac{1}{3} \leqq z \leqq \frac{2}{3}, x=y=0\right\}.$$

B^1 in Figure 3.13 is

$$\left\{(t,x,y,z)|t=0, 0 \leqq z \leqq \frac{1}{3}, \frac{2}{3} \leqq z \leqq 1, x=y=0\right\}$$
$$\cup \left\{(t,x,y,z)|-0.5 \leqq t \leqq 0, z=\frac{1}{3}, z=\frac{2}{3}, x=y=0\right\}$$
$$\cup \left\{(t,x,y,z)|t=-0.5, \frac{1}{3} \leqq z \leqq \frac{2}{3}, x=y=0\right\}.$$

While it is possible to write the literal expressions shown above, we highly recommend gathering intuition from conceptual figures. It is not easy to form a concept of 4-space from expressions alone. However, by first *seeing* objects in 4-space, intuition can follow, and one may write explicit expressions much more easily.

3.3 Ribbon-Moves on a Disjoint Union of a Sphere and a Torus in $\mathbb{R}^4$

The object pictured in Figure 3.14 is called the *torus*, denoted by T^2. The torus is like an inner tube, or the outer surface of a donut.

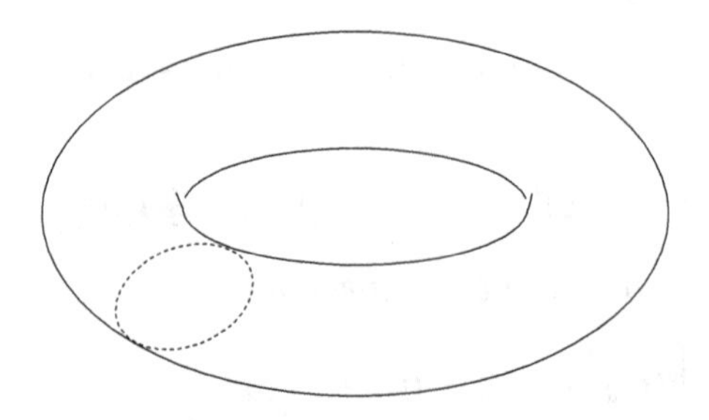

Figure 3.14 Torus.

If the interior is filled, then it is called the *solid torus.*

In Figure 3.15, we show S^2 and T^2 embedded in $\mathbb{R}^3$.

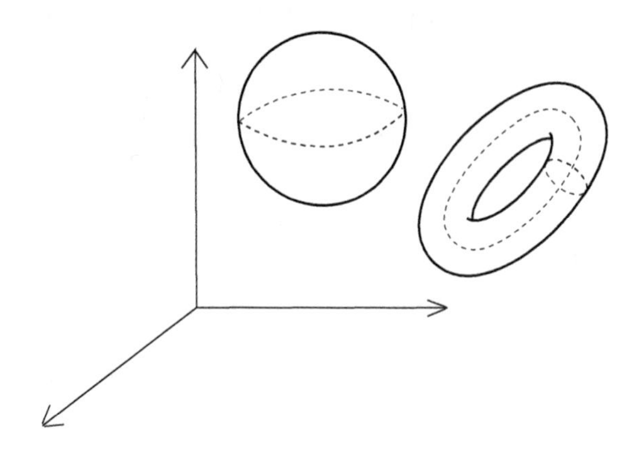

Figure 3.15 S^2 and T^2 in $\mathbb{R}^3$.

Now consider a disjoint union $S^2 \amalg T^2$ of S^2 and T^2 in $\mathbb{R}^4$.

We construct two types of disjoint unions $S^2 \amalg T^2$ in $\mathbb{R}^4$ — call them L_0 and L_1 — such that neither T^2 nor S^2 touches itself.

L_0: S^2 bounds a three-dimensional ball B^3 embedded in $\mathbb{R}^4$.
T^2 bounds the solid torus embedded in $\mathbb{R}^4$.
B^3 does not touch the solid torus.

We can easily construct L_0. Regarding $\mathbb{R}^4$ as the $xyzt$-space and $\mathbb{R}^3$ as the xyz-space in the $xyzt$-space, we show the construction in Figure 3.16.

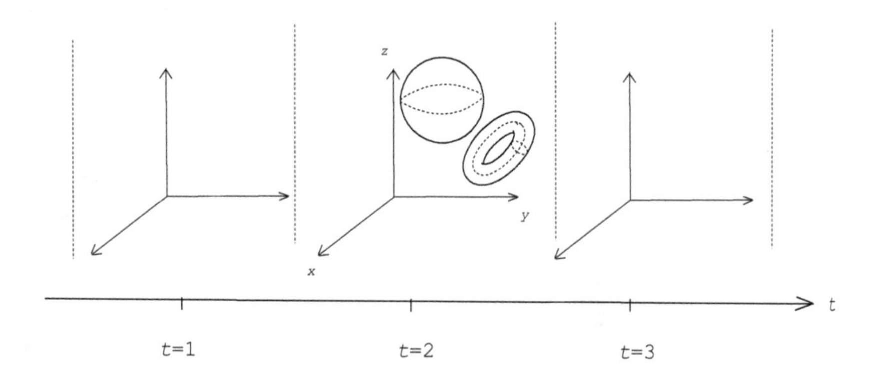

Figure 3.16 L_0.

L_1: S^2 bounds a three-dimensional ball B^3 embedded in $\mathbb{R}^4$, where we do not care whether or not B^3 intersects T^2.

T^2 bounds the solid torus embedded in $\mathbb{R}^4$ so that the solid torus does not touch S^2.

No matter how we draw B^3, it intersects the solid torus. But, we can draw B^3 and the solid torus so that their intersection is a single 2-ball.

Perhaps, the reader is thinking that L_1 exists in this case, too. And that is true!

Can you imagine L_1? Can you construct L_1? To construct it, you simply have to replace S^1 in Figure 3.8 with a solid torus.

But before we draw this construction of L_1 in $\mathbb{R}^4$, we go through another construction of L_0 in $\mathbb{R}^4$ — different from that shown in Figure 3.16.

Put S^2 and T^2 in $\mathbb{R}^4$ as shown in Figure 3.17.

T^2 exists in $1 \leqq t \leqq 3$:

$T^2 \cap \mathbb{R}^3$ at $t = 1$ is an annulus.

$T^2 \cap \mathbb{R}^3$ at each t of $1 < t < 3$ is two circles.

$T^2 \cap \mathbb{R}^3$ at $t = 3$ is an annulus.

Although this torus has "corners," we still regard it as T^2 for the same reason as we would regard a rectangle as S^1, or a stuffed cylinder as B^3: we can smoothly transform our torus-with-corners without touching it, into what we typically picture as T^2. Note that in Figure 3.17, S^2 and T^2 make L_0.

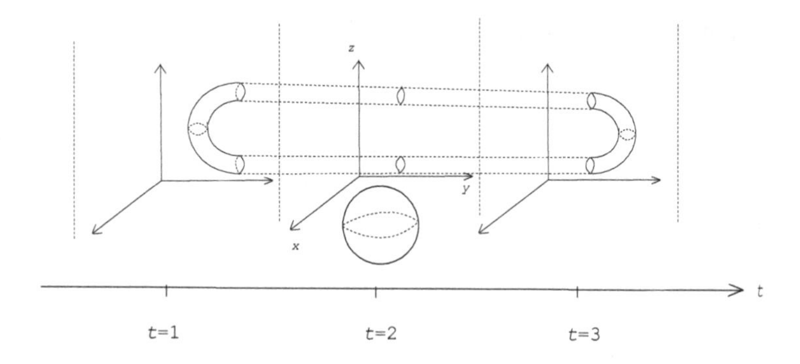

Figure 3.17 L_0.

We now construct L_1 as follows:

Embed S^2 and T^2 in $\mathbb{R}^4$ as shown in Figure 3.18, placing T^2 in the same position as shown in Figure 3.17. $T^2 \cap (\mathbb{R}^3$ at $t = 2)$ is two circles, one of which exists in the inside of S^2 in $\mathbb{R}^3$ at $t = 2$, and the other in the outside. Recall that in $\mathbb{R}^2$, S^1 bounds a disk, separating the inside of the circle from the outside. In $\mathbb{R}^3$, that is not the case — the inside of the circle is connected to the outside. Similarly, when we embed S^2 in $\mathbb{R}^4$, there is no separation between the part of space inside the sphere and the part of space that is outside. In Figure 3.18, S^2 and T^2 as described above make L_1.

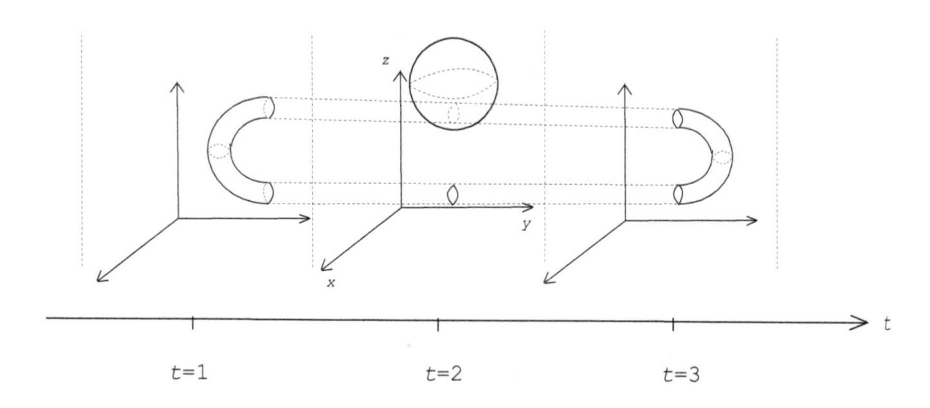

Figure 3.18 L_1.

It is known in this case, too, that L_0 and L_1 are different.

Consider an analogy of the case of $S^1 \amalg S^2$ in $\mathbb{R}^4$ outlined in Section 3.2. L_1 is changed into L_0 by an operation (a *local move*) as drawn in Figure 3.19. The procedure in this case is drawn in Figure 3.20. An *annulus*, as written in Figure 3.19, is the side of a cylinder.

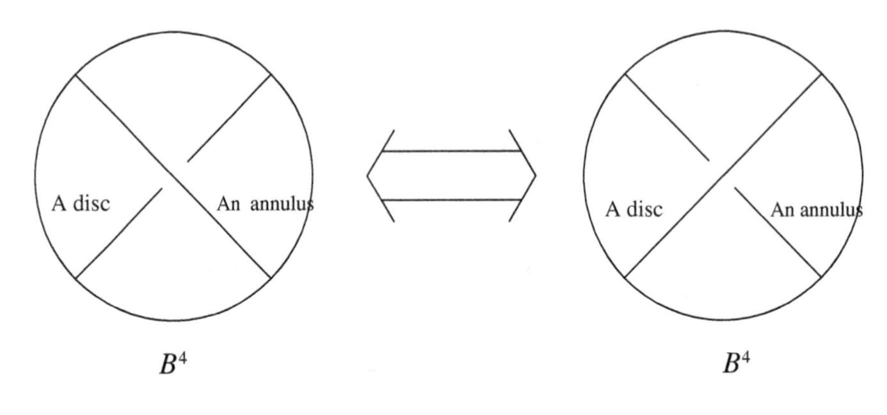

$$S^2 \cap B^4 = \text{A disc}$$
$$T^2 \cap B^4 = \text{An annulus}$$

Figure 3.19 The ribbon-move.

This procedure is also called the *ribbon-move*. We say that L_1 (respectively, L_0) is changed into L_0 (respectively, L_1) by one ribbon-move. Figures 3.21 and 3.22 show ribbon-moves more explicitly.

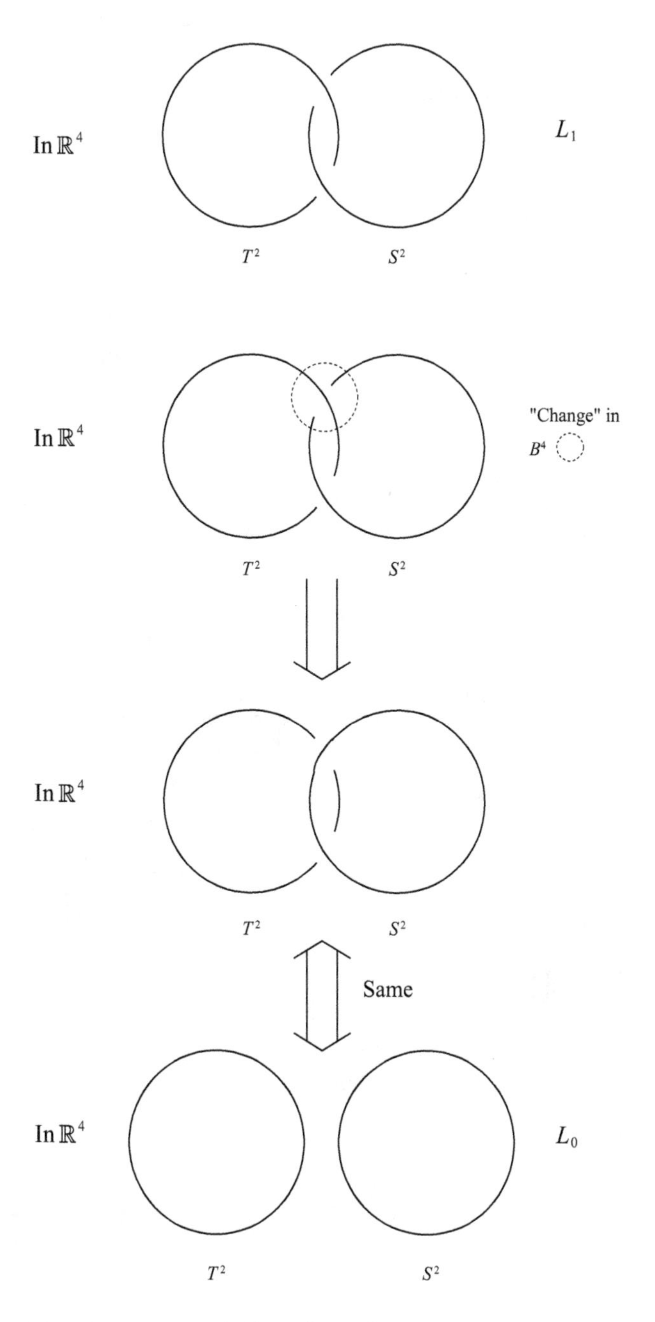

Figure 3.20 A sequence of $S^2 \amalg T^2$ in $\mathbb{R}^4$ generated by the ribbon-move.

The method illuminated in these figures is analogous to that shown in Figures 3.12 and 3.13.

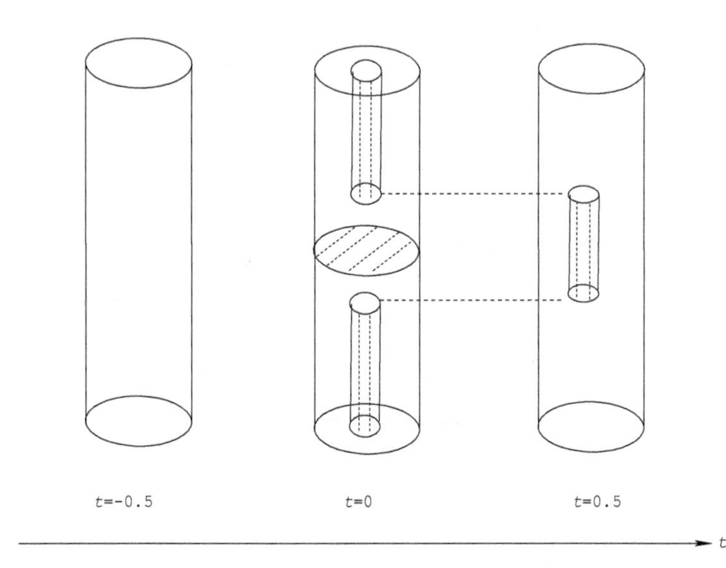

Figure 3.21 The ribbon-move. This figure continues in Figure 3.22.

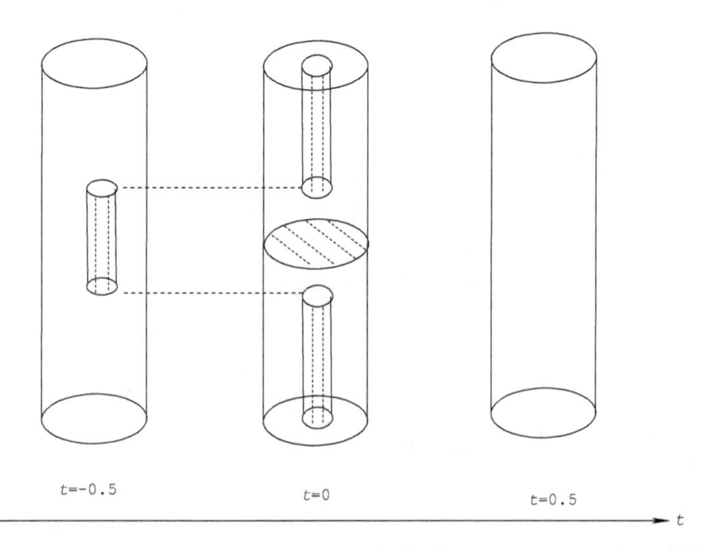

Figure 3.22 The ribbon-move. This figure is a continuation of Figure 3.21.

Chapter 4

Knotted-Objects in 4-Space and Beyond

4.1 Ribbon-Moves on Two-Dimensional Knots

If a single copy of a sphere S^2 is embedded in $\mathbb{R}^4$, it is called a *two-dimensional knot* or *2-knot*. If a set of m copies of S^2 is embedded in $\mathbb{R}^4$, it is called an *m-component 2-link*.

A circle S^1 embedded in $\mathbb{R}^3$ is called a knot in the previous sections. It is sometimes called a *one-dimensional knot* or *1-knot*.

If a 2-knot K bounds a 3-ball B^3 embedded in $\mathbb{R}^4$, we call K the *trivial 2-knot*. The *trivial m-component 2-link* bounds a disjoint union of m copies of B^3.

There does exist a trivial 2-knot in 4-space. Take S^2 in $\mathbb{R}^3$ such that S^2 bounds a 3-ball $B^3 \subset \mathbb{R}^3$. Regard $\mathbb{R}^3$ as $\mathbb{R}^3 \times \{0\}$ in $\mathbb{R}^4$. We obtain S^2 in $\mathbb{R}^4$. This is the trivial 2-knot.

You may ask: Is there a non-trivial 2-knot? There is! We show an example in what follows, in this section.

We will write another explicit method to construct non-trivial 2-knots in Section 5.1.2 of Chapter 5. Read both. Which do you prefer?

We introduce the *ribbon-move* on two-dimensional knots in $\mathbb{R}^4$, and make use of them to construct a non-trivial 2-knot.

The ribbon-move on a 2-knot can be defined in the same manner as that in the case of $S^2 \amalg T^2$ in $\mathbb{R}^4$. Recall B^4 in $\mathbb{R}^4$ as shown in Figures 3.9, 3.11, and 3.12. The intersection of the 2-knot and B^4 is a disjoint union of a disc and an annulus. Take a 4-ball B^4 in $\mathbb{R}^4$ and assume that a 2-knot K and B^4 intersect as shown in Figure 3.11. Change the part of K shown in Figure 3.11 into that shown in Figure 3.12.

Before we construct our non-trivial 2-knot, we show a sequence of 1-knots in Figure 4.1.

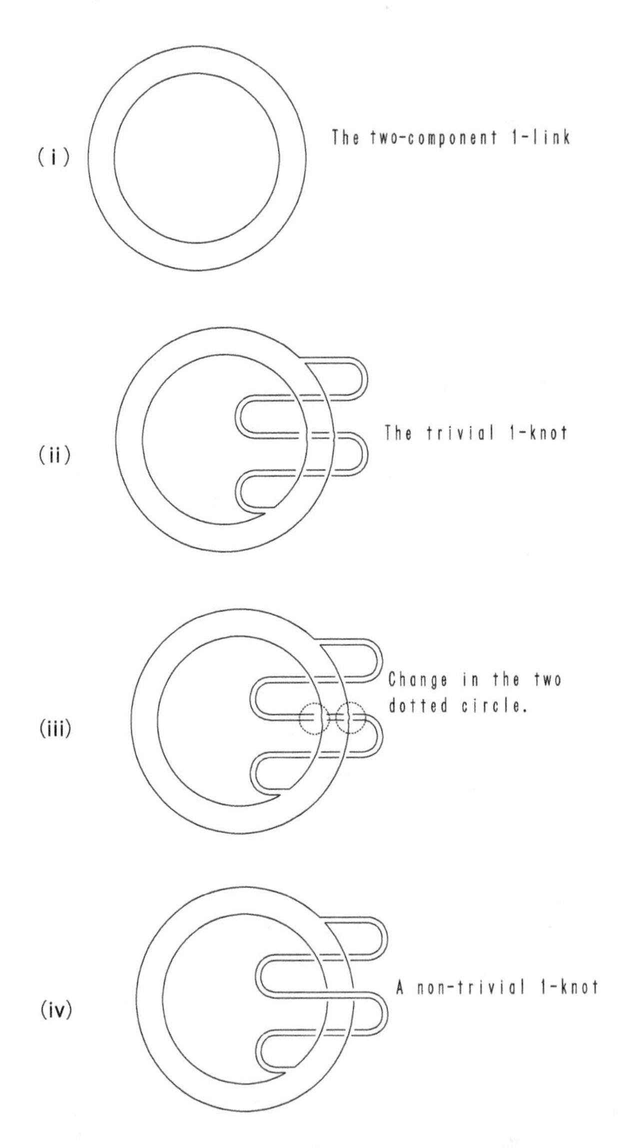

Figure 4.1 This is a picture in $\mathbb{R}^3$. (ii)–(iv) is a sequence of 1-knots.

In Figure 4.6, we generalize Figure 4.1 by showing a sequence of a finite number of ribbon-moves which make the trivial 2-knot into a non-trivial 2-knot.

We call the object in Figure 4.2 a tube. We may transform a tube smoothly without touching it. We can transform a *tube* smoothly without touching it, into an annulus. Note that the boundary of the tube is two circles.

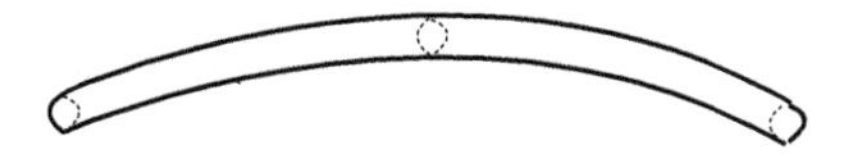

Figure 4.2 A tube.

Take two spheres in $\mathbb{R}^3$ as shown in Figure 4.3.

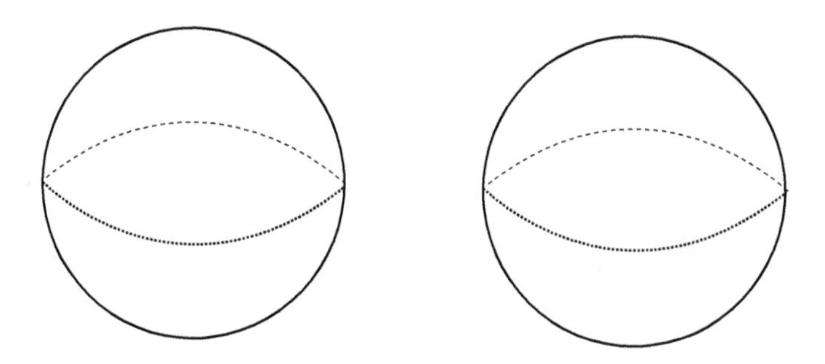

Figure 4.3 Two spheres.

Remove a small disc from each sphere in Figure 4.3, and connect by using a tube as in Figure 4.4. The result is one sphere since we can transform this object smoothly without touching it, into the sphere.

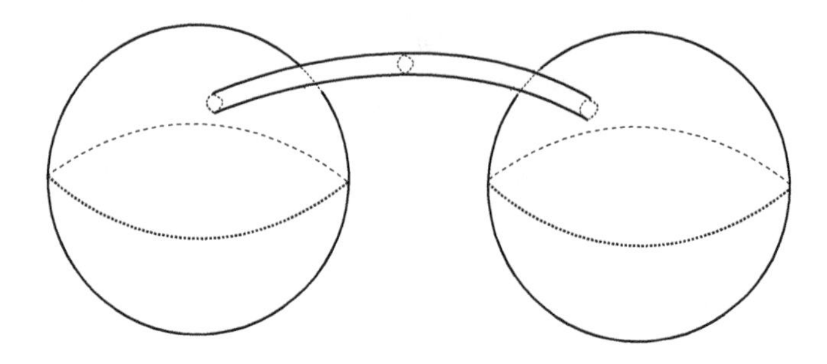

Figure 4.4 One sphere made from two and a tube.

Put two spheres in $\mathbb{R}^3$ at $t = 0$ in $\mathbb{R}^4$ as drawn in Figure 4.5.

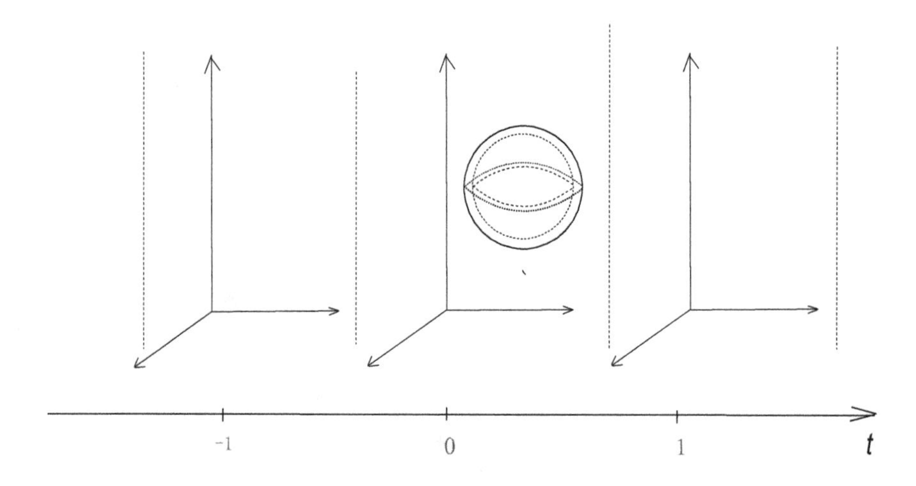

Figure 4.5 Two spheres.

We call a tube without its boundary the *interior* of the tube. Put the interior of the tube in the $t > 0$ region of $\mathbb{R}^4$ in Figure 4.5. Do the operation in Figure 4.4. We get one sphere in the $t \geqq 0$ region of $\mathbb{R}^4$, not at $t = 0$ of $\mathbb{R}^4$. We do this when we change Figure 4.6(i) into Figure 4.6(ii), though more abstractly. Try to imagine explicitly our tube interior in the $t > 0$ region!

We now show an example of changing the trivial 2-knot into a non-trivial 2-knot in Figure 4.6. This operation is very important! Picture the operation and the resulting non-trivial knot in 4-space.

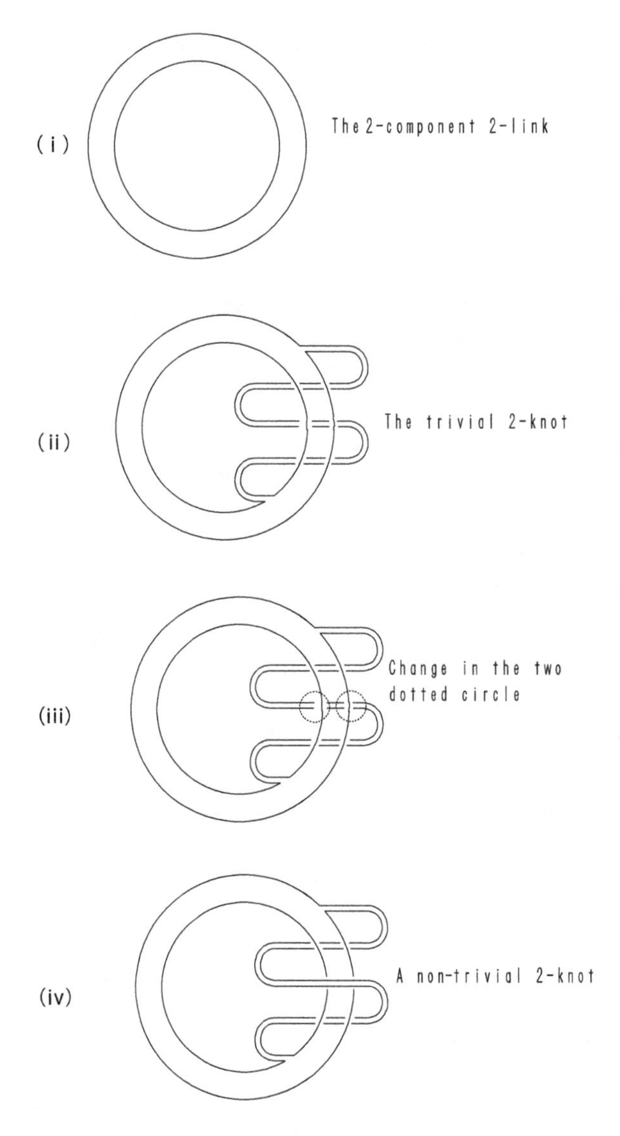

Figure 4.6 This is a picture in $\mathbb{R}^4$. We carry out the ribbon-move in (iii).

Figure 4.6(i) represents two spheres in $\mathbb{R}^3$ at $t = 0$ in $\mathbb{R}^4$ of Figure 4.5.

Change the two spheres shown in Figure 4.6(i) into the singular sphere shown in Figure 4.6(ii) by using a tube. In Figure 4.6(ii), put the interior of the tube in the $t > 0$ region of 4-space. The one 2-sphere Figure 4.6(ii) in $\mathbb{R}^4$ is the trivial 2-knot in $\mathbb{R}^4$.

Carry out the ribbon-moves twice on this 2-knot in the dotted circle of Figure 4.6(iii), each of which represents a small three-dimensional ball. Some parts of the tube go to the region $t \leqq 0$.

The result is drawn in Figure 4.6(iv). It is a non-trivial 2-knot. Some parts of the knot exist in $t > 0$, some parts in $t < 0$, and the other parts at $t = 0$.

We have thus changed the trivial 2-knot in Figure 4.6(ii) into the non-trivial 2-knot shown in Figure 4.6(iv).

For more details, see the author's papers [85, 88, 90, 93, 95–97].

Note: If K_1 is obtained from K_2 by one ribbon-move, then we also say that K_2 is obtained from K_1 by one ribbon-move.

Two 2-links K_1 and K_2 are said to be *ribbon-move equivalent* if there are 2-links $K_1 = \bar{K}_1, \bar{K}_2, \ldots, \bar{K}_{r-1}, \bar{K}_r = K_2$ (r is a natural number) such that $\bar{K}_i$ is obtained from $\bar{K}_{i-1}$ $(1 < i \leqq r)$ by one ribbon-move.

The following is known:

Any m-component one-dimensional link is made into the trivial m-component one-dimensional link by a sequence of a finite number of crossing changes, where m is any natural number.

In particular, any one-dimensional knot is made into the trivial one-dimensional knot by a sequence of a finite number of crossing changes.

Similarly, we can make countably infinitely many non-trivial 2-knots in $\mathbb{R}^4$ from the trivial 2-knot in $\mathbb{R}^4$ by a sequence of a finite number of ribbon-moves. An example of this is illustrated in Figure 4.6.

It is natural to ask the following question.

Question 4.1. Are all two-dimensional knots in $\mathbb{R}^4$ ribbon-move equivalent to the trivial two-dimensional knot in $\mathbb{R}^4$?

The author shows in [85] that the answer is in fact negative. We go through this result in Section 5.2 of Chapter 5.

Furthermore, the author proves that there are countably infinitely many non-trivial 2-knots which are not ribbon-move equivalent to the trivial knot. His papers [90, 93, 95–97] are sequels of the paper cited above.

It is well known that there is a 1-knot which can be made into the unknot by two crossing changes, but which cannot be changed into the unknot by one crossing change. It is thus natural to ask the following question.

Question 4.2. Is there a 2-knot which is made into the trivial 2-knot by two ribbon-moves, but which cannot be changed into the trivial 2-knot by one ribbon-move?

In [97], the author gives an affirmative answer to Question 4.2. See Section 5.3 of Chapter 5.

Let K and K' be arbitrary 2-knots. Then is K ribbon-move equivalent to K'? That is, classify 2-knots by the ribbon-move equivalence. This is an open problem.

The following is also an open question: Is there a type of local move on 2-knots which can unknot all 2-knots?

4.2 High-Dimensional Knots

Let's move into higher-dimensional space!

We explain local moves on high-dimensional knots. High-dimensional knot theory is a branch of knot theory — an area of topology — which is connected to many other fields of mathematics and physics and so on.

We define *n-dimensional space* $\mathbb{R}^n$ as the set $\{(x_1, \ldots, x_n)\}$ of all ordered n-tuples $(x_1, \ldots, x_n)$ of real numbers. We may also write $\{(x_1, \ldots, x_n) | x_i \text{ is a real number}\}$.

We define *n-dimensional ball*, or *n-ball* B^n, as follows: $\{(x_1, \ldots, x_n) | x_1^2 + \cdots + x_n^2 \leqq 1\}$. Note that B^1 is a line segment and B^2 is a disc.

The *n-dimensional sphere*, or *n-sphere* S^n, is $\{(x_1, \ldots, x_n) | x_1^2 + \cdots + x_n^2 = 1\}$. Note that S^1 is a circle and S^2 is a sphere. Moreover, note that the boundary of the n-ball B^n is S^{n-1} (the *boundary* is a mathematical term). Consider S^0. This set will be two points — more precisely, these points are the boundary of a line segment.

If one n-sphere S^n is embedded in $\mathbb{R}^{n+2}$, it is called an *n-dimensional knot* or an *n-knot*. If a set of m copies of S^n is embedded in $\mathbb{R}^{n+2}$, it is called an *m-component* $(n+2)$*-link*.

If an n-knot K bounds an $(n + 1)$-ball B^{n+1} embedded in $\mathbb{R}^{n+2}$, we call K the *trivial n-knot.* The *trivial m-component n-link* bounds a disjoint union of m copies of B^{n+1}.

You may be wondering: Is there a non-trivial high-dimensional knot? There is!

We provide examples in what follows, in Sections 4.3 and 4.4, obtained by introducing two kinds of local moves on high-dimensional knots. We introduce generalizations to higher dimensions of the pass-move and the crossing change on 1-knots.

We will describe another explicit method to construct non-trivial high-dimensional knots in Section 5.1.3 of Chapter 5. Read these three ways. Which do you prefer?

4.3 High-Dimensional Pass-Moves

Recall the pass-move on 1-links. Figure 4.7 provides an illustration of the pass-move.

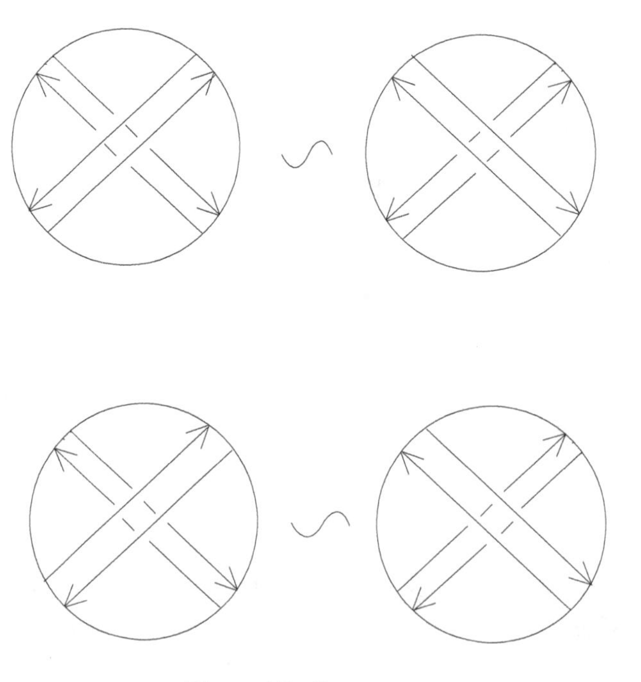

Figure 4.7 Pass-move.

We review an example of a pass-move that changes the trivial knot into a non-trivial knot in Figures 4.8–4.10.

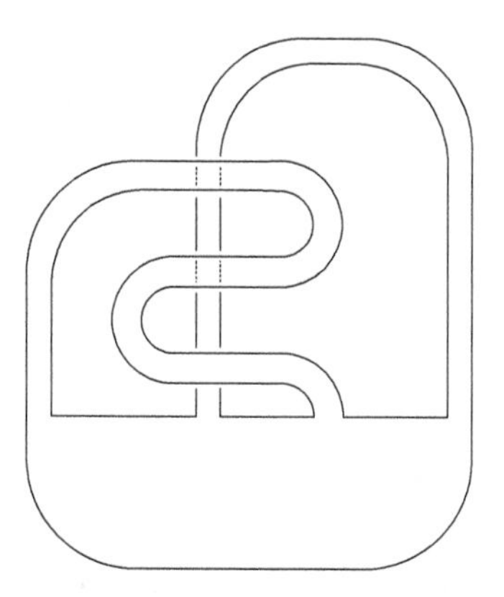

Figure 4.8 Note that this is the one-component trivial link, or the trivial knot.

The 1-knot in Figure 4.8 is the trivial 1-knot.

Now you try: Carry out a pass-move on this 1-knot in of Figure 4.9.

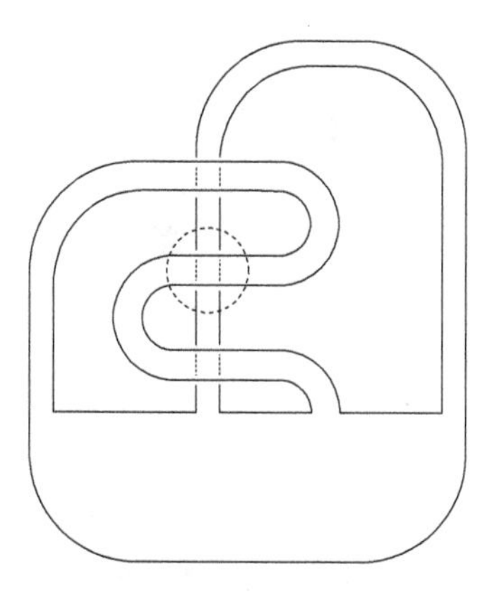

Figure 4.9 This knot is the same as that in Figure 4.8.

The resulting knot shown in Figure 4.10 is a non-trivial 1-knot.

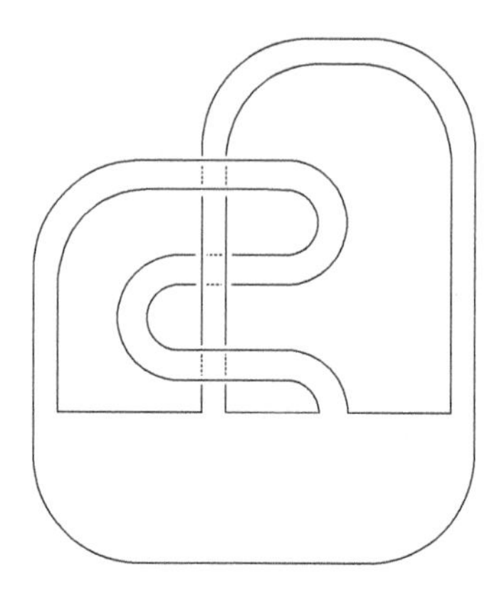

Figure 4.10 A pass-move makes this knot from that in Figure 4.9 by a ribbon-move.

In the following sections, we introduce a notation $\times$, then go on to generalize the pass-move operation to higher dimensions.

Let A be in $\mathbb{R}^n$. Let B be in $\mathbb{R}^m$. Define $A \times B$ to be $\{(x_1, \ldots, x_n, y_1, \ldots, y_m) \mid (x_1, \ldots, x_n)$ is in A; $(y_1, \ldots, y_m)$ is in $B\}$ in $\mathbb{R}^{n+m}$. See Figure 4.11.

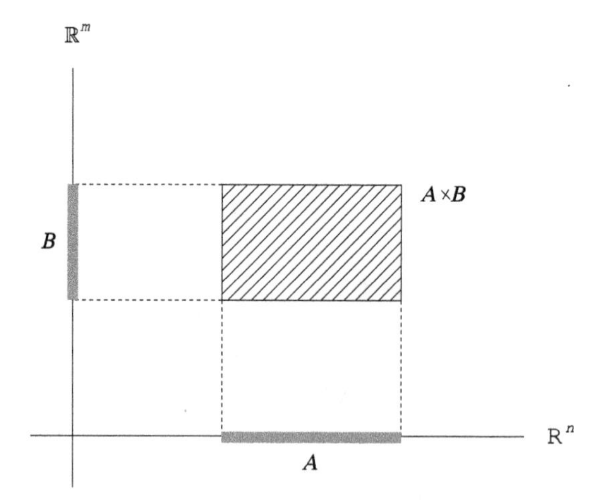

Figure 4.11 $A \times B$.

$S^1 \times S^1$ is an example. S^1 is represented by

$$\{(x_1, x_2)|(x_1)^2 + (x_2)^2 = 1\}$$

in $\mathbb{R}^2$, and $S^1 \times S^1$ is represented by

$$\{(x_1, x_2, y_1, y_2) | (x_1)^2 + (x_2)^2 = 1, \quad (y_1)^2 + (y_2)^2 = 1\}$$

in $\mathbb{R}^4$.

Note: We have defined $S^1 \times S^1$ in $\mathbb{R}^4$, but it can be moved into without touching it. and embedded in $\mathbb{R}^3$. $S^1 \times S^1$ yields the torus, which we defined in Section 3.3 of Chapter 3. It is a good exercise to imagine this movement.

See Figure 4.12. You may understand why $T^2 = S^1 \times S^1$. A circle S^1 runs along another circle S^1. The trace is the torus T^2.

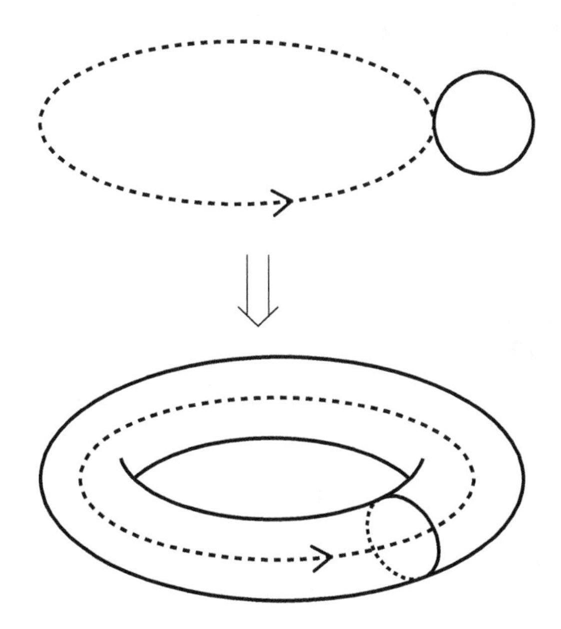

Figure 4.12 The trace of S^1 running along S^1 is T^2.

Regard S^1 as the union of two 1-balls B^1_u and B^1_d. Recall that the 1-ball is a segment. Then $S^1 \times S^1$ is regarded as the union of four parts, $B^1_u \times B^1_u$, $B^1_u \times B^1_d$, $B^1_d \times B^1_u$, and $B^1_d \times B^1_d$. Note that $S^1 \times S^1$ is the torus.

Remove the interior of $B^1_u \times B^1_u$ from $S^1 \times S^1$ — call it F. F is drawn abstractly as shown in Figure 4.13.

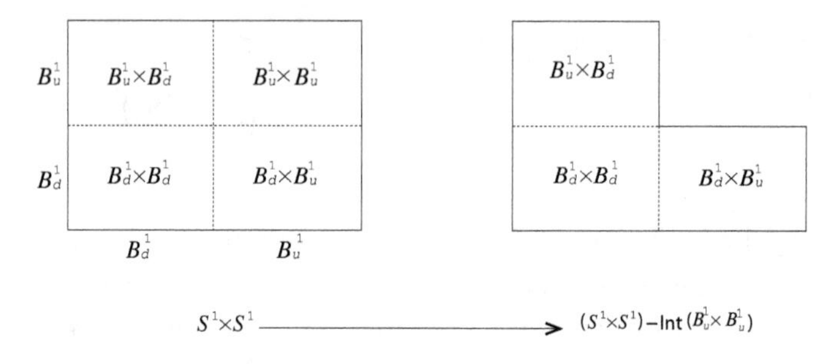

Figure 4.13 F.

We draw F explicitly in Figure 4.14. Note that we can smooth out the corner of $B^1_u \times B^1_u$ to change it into the 2-ball. Hence, the boundary of F is a single circle.

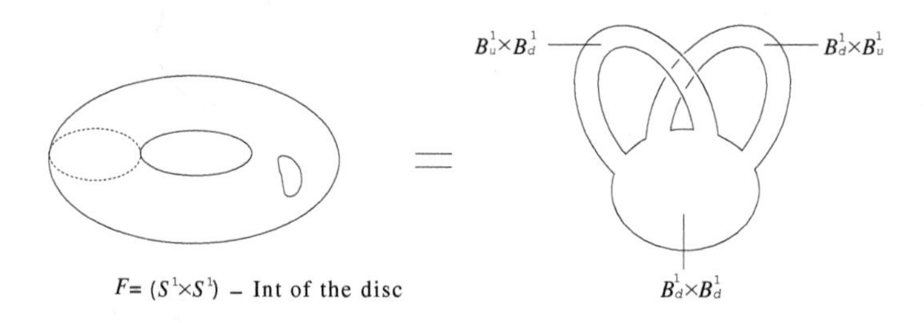

Figure 4.14 F.

We carry out a pass-move on a 1-knot K that bounds F below. Take F in $\mathbb{R}^3$ as drawn in Figure 4.15.

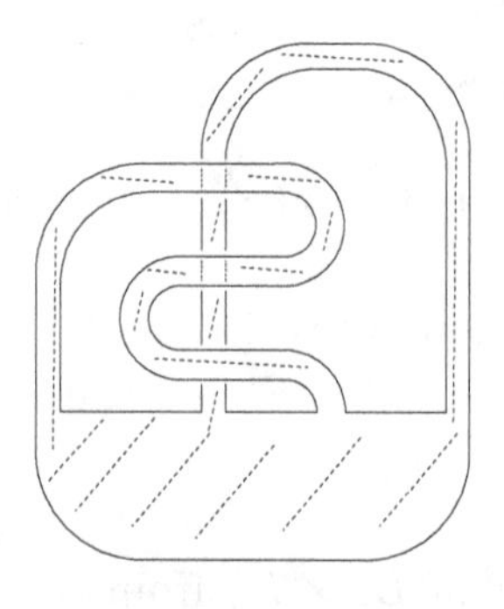

Figure 4.15 The boundary of F is the trivial knot.

The boundary of F in $\mathbb{R}^3$ is a 1-knot, and more precisely, the trivial knot.

Carry out a pass-move on this knot in ◯ of Figure 4.16.

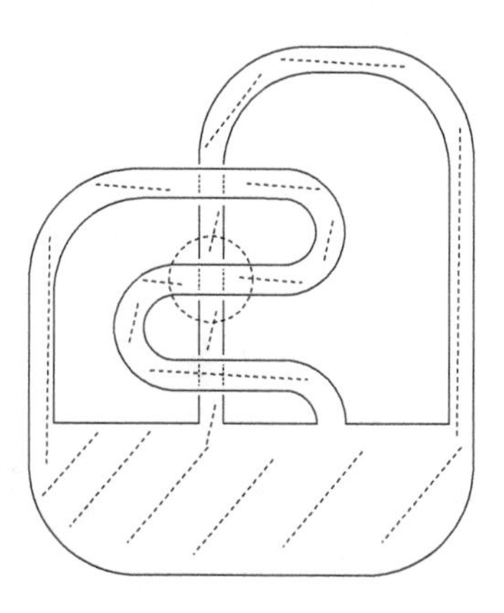

Figure 4.16 This figure is the same as in Figure 4.15 except for the dotted circle, which represents the $(n+2)$-ball.

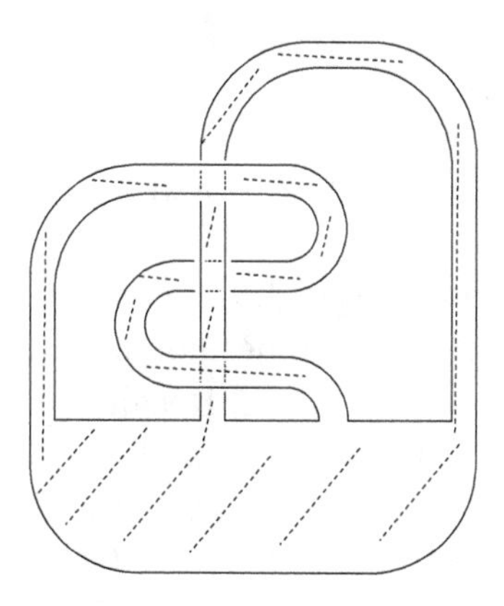

Figure 4.17 The boundary of F is a non-trivial knot.

The resulting knot in Figure 4.17 is a non-trivial knot.

Note that the pass-move is carried out with a surface, as drawn in Figure 4.18.

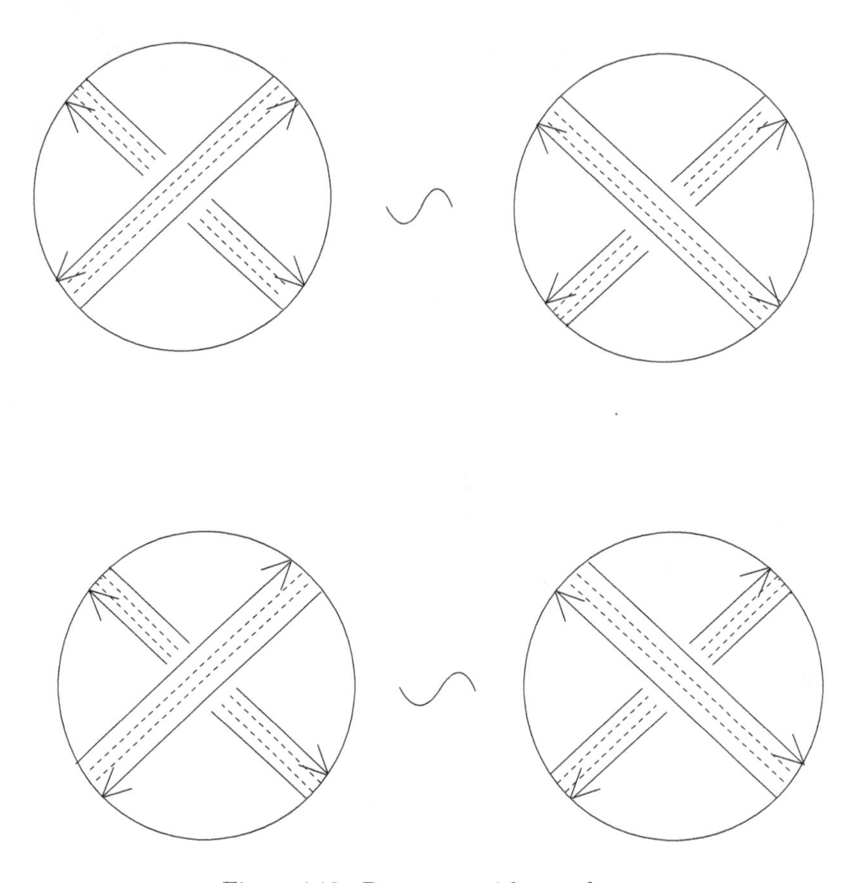

Figure 4.18 Pass-move with a surface.

We introduce a non-trivial high-dimensional knot in what follows.

We change the trivial n-dimensional knot into a non-trivial n-dimensional knot by a local move. Recall that n-knots are in $\mathbb{R}^{n+2}$.

Let $p + q = n + 1$. Regard S^p as the union of p-balls B^p_u and B^p_d, and S^q as the union of the q-balls B^q_u and B^q_d. Then $S^p \times S^q$ is regarded as the union of four parts, $B^p_u \times B^q_u$, $B^p_u \times B^q_d$, $B^p_d \times B^q_u$, and $B^p_d \times B^q_d$. Remove the interior of $B^p_u \times B^q_u$ from $S^p \times S^q$ — call it F. F is drawn in Figure 4.19.

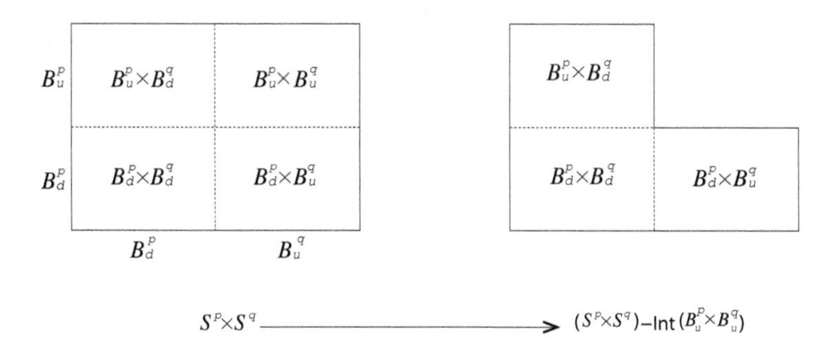

Figure 4.19 F.

F is drawn in another way as shown in Figure 4.20. Recall that we can smooth out the corner of $B^p_u \times B^q_u$ and deform it into the $(p+q)$-dimensional ball. Since $p+q=n+1$, the boundary of F is S^n.

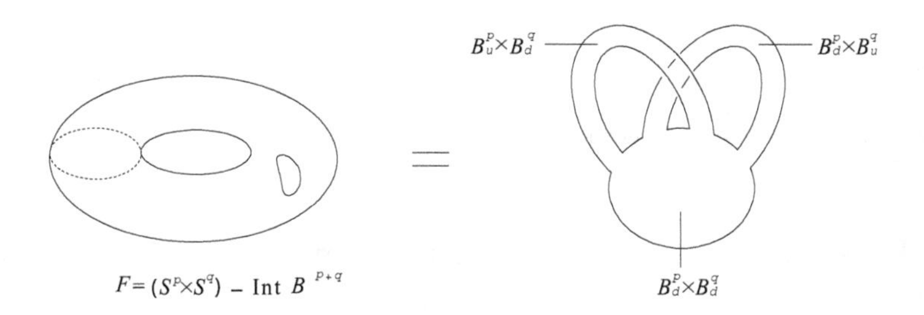

Figure 4.20 F.

Take F in $\mathbb{R}^{n+2}$ as shown in Figure 4.21.

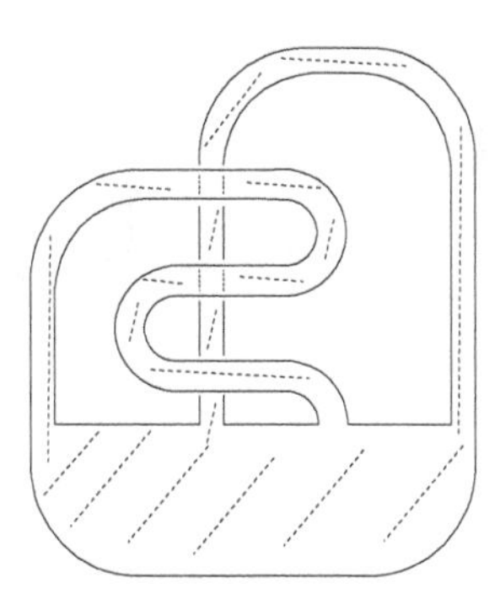

Figure 4.21 The trivial n-knot in $\mathbb{R}^{n+2}$.

The boundary of F in $\mathbb{R}^{n+2}$ is an n-knot — specifically the trivial n-knot.

Carry out a local move on this n-knot in ◯ of Figure 4.22.

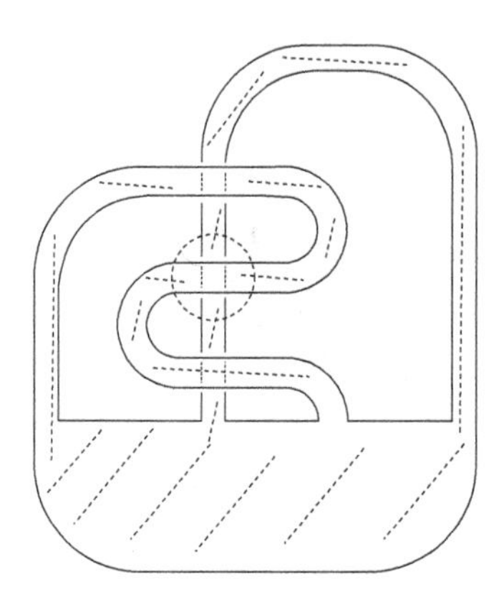

Figure 4.22 This n knot is the same as in Figure 4.21 except for a dotted circle, which represents the $(n+2)$-ball.

We see that the resulting n-knot in Figure 4.23 is a non-trivial n-knot.

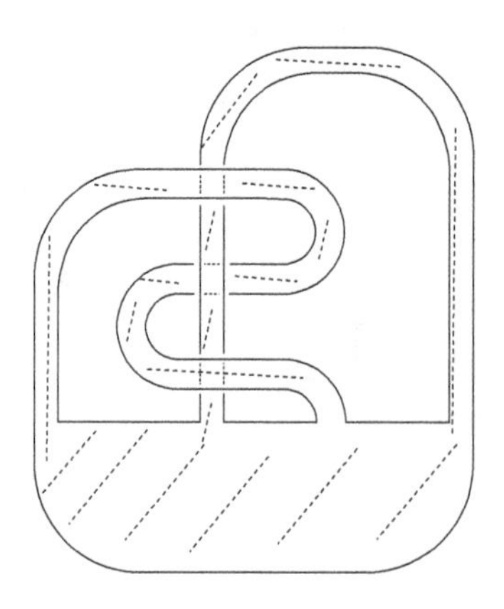

Figure 4.23 A non-trivial n-knot.

We call the above local move the (p, q)-*pass-move* on n-dimensional knots (for $p+q=n+1$ and $n \geqq 1$). More precisely, we define as in the following paragraphs. See the author's papers [79, 85, 87, 90, 93, 95, 97] for further reading.

Take an $(n+2)$-ball B^{n+2} trivially embedded in $\mathbb{R}^{n+2}$ and regard B^{n+2} as $B^1 \times B^p \times B^q$. Note that B^l is $\{(x_1, \ldots, x_l)|(x_1)^2 + \cdots + (x_l)^2 \leqq 1\}$ for any natural number l. In particular, B^1 is $[-1, 1] = \{x| -1 \leqq x \leqq 1\}$.

Now, consider a smaller p-ball D^p in B^p and suppose that D^p is $\{(x_1, \ldots, x_p)|(x_1)^2 + \cdots + (x_p)^2 \leqq \frac{1}{4}\}$. Consider a smaller q-ball D^q in B^q as well.

Take $D^p \times B^q$ and $B^p \times D^q$ in the left (respectively, right) figure of B^{n+2} as shown in Figure 4.24, and denote the union of $D^p \times B^q$ and $B^p \times D^q$ as U_+ (respectively, U_-). Note that $D^p \times B^q$ and $B^p \times D^q$ do not touch each other.

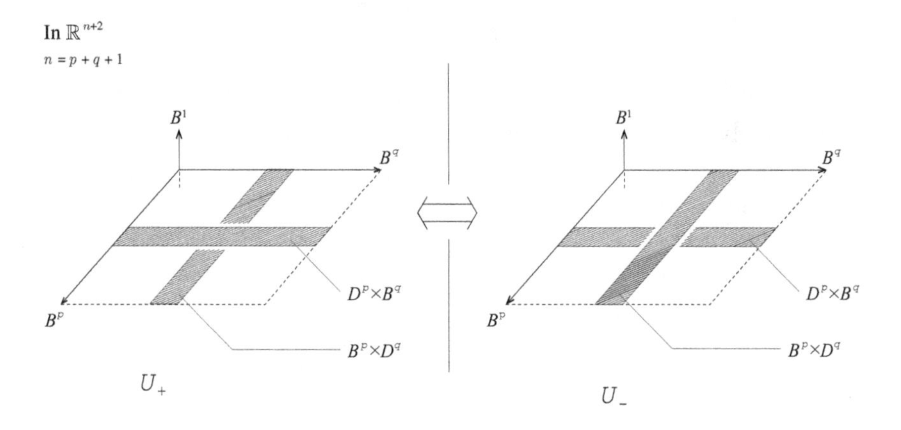

Figure 4.24 The (p, q)-pass-move.

$B^p \times D^q$ of U_+ (respectively, U_-) is embedded in $\{0\} \times B^p \times B^q$, $D^p \times B^q$ of U_+ is embedded in $\{x \geqq 0\} \times B^p \times B^q$, and $D^p \times B^q$ of U_- is embedded in $\{x \leqq 0\} \times B^p \times B^q$; all of these embeddings are trivial.

S^{p-1} is the boundary of D^p. Hence, $S^{p-1} \times B^q$ is included in the boundary of $D^p \times B^q$, and $B^p \times S^{q-1}$ is included in the boundary of $B^p \times D^q$.

Take $S^{p-1} \times B^q$ and $B^p \times S^{q-1}$ in the left (respectively, right) figure of B^{n+2} as shown in Figure 4.25, and denote the union of $S^{p-1} \times B^q$ and $B^p \times S^{q-1}$ as V_+ (respectively, V_-).

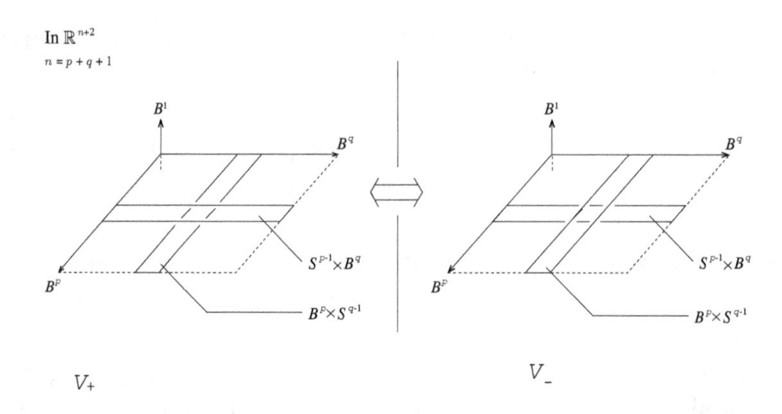

Figure 4.25 The (p, q)-pass-move.

Figures 4.26 and 4.27 are more concrete representations of the (p, q)-pass-move.

Let K_+ and K_- be n-dimensional knots in $\mathbb{R}^{n+2}$. Suppose that K_+ and K_- differ only in the above B^{n+2}, and that the intersection of K_+ (respectively, K_-) and B^{n+2} is precisely V_+ (respectively, V_-). Then we say that K_- (respectively, K_+) is obtained from K_+ (respectively, K_-) by one *(p, q)-pass-move.*

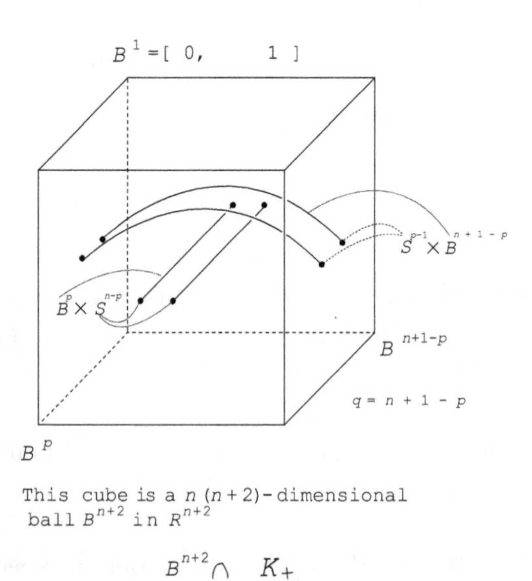

Figure 4.26 The (p, q)-pass-move. This figure continues in Figure 4.27.

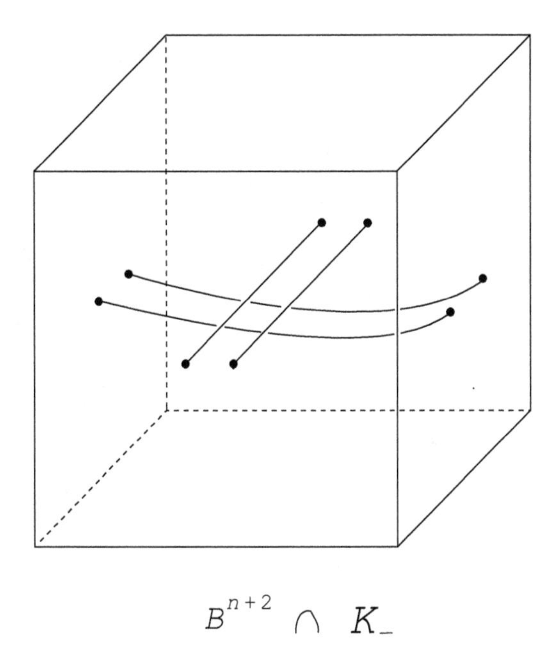

Figure 4.27 The (p, q)-pass-move. This figure is a continuation of Figure 4.26.

In order to imagine the higher-dimensional case, it is very often useful to consider a lower-dimensional case. Consider the case where $p = q = 1$ and hence $n = 3$. This case — that is, the $(1, 1)$-pass-move — is equivalent to the ordinary pass-move we defined on 1-knots.

Recall that S^0 is two points. Compare Figure 4.24 with Figure 4.18, Figure 4.25 with Figure 4.7, the pair of Figures 4.26 and 4.27 with Figure 4.7.

Next consider the case where $p = 1$, $q = 2$, and $n = 4$. You may note that one $(1, 2)$-pass-move is two ribbon-moves (Figures 4.28 and 4.29).

Note that the operation in Figure 4.6 is the same as the $(p, q) = (1, 2)$ case of the operation from Figure 4.22 to Figure 4.23.

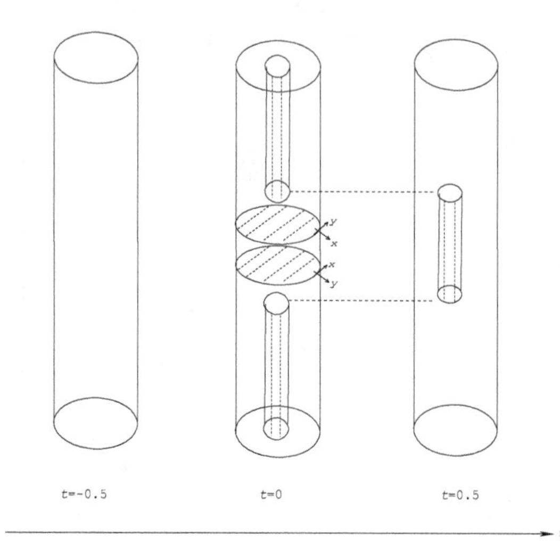

Figure 4.28 The $(1, 2)$-pass-move. This figure continues in Figure 4.29.

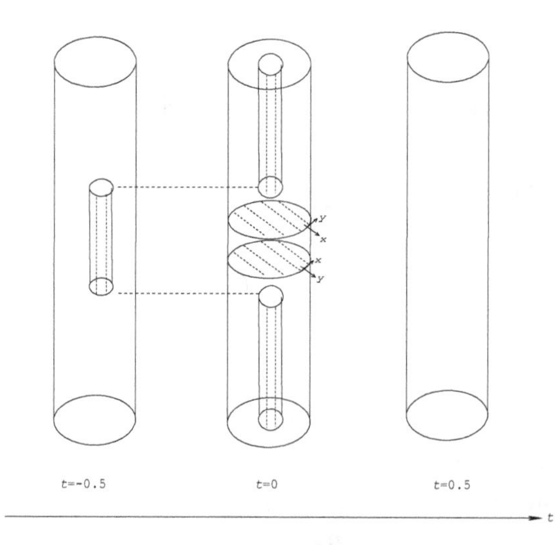

Figure 4.29 The $(1, 2)$-pass-move. This figure is a continuation of Figure 4.28.

As we showed in Figures 4.21–4.23, we can obtain a non-trivial n-knot from the trivial n-knot by one (p, q)-pass-move, where $p+q = n+1$.

It is known that there are countably infinitely many non-trivial n-knots which are (p, q)-pass-move equivalent to the trivial n-knot.

However, it is not completely known what kinds of high-dimensional knots are (p,q)-pass-move equivalent to the trivial knot — both in the case of fixing p and q and in the case of fluctuating p and q. The case $p = q \geqq 2$ is solved by the author using [44, Proposition 12.1] and [79, Theorem 4.1].

4.4 Twist-Moves on $(2m+1)$-Dimensional Knots

We introduce another local move on high-dimensional knots called the *twist-move*. We define the twist-move on $(2m+1)$-dimensional knots in $\mathbb{R}^{2m+3}$.

Before that, though, we review the crossing change on 1-links and then look at a generalization to higher dimensions. This will help us define the twist-move on high-dimensional knots.

Figure 4.30 shows a crossing change on 1-links, as defined in Section 1.2 of Chapter 1.

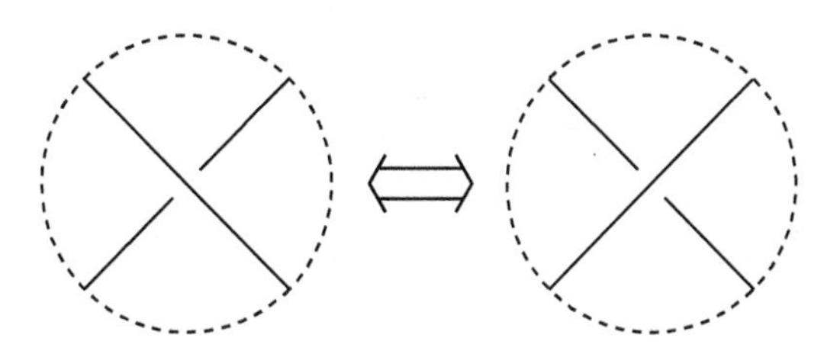

Figure 4.30 Crossing change.

Now consider the local move drawn in Figure 4.31. The result is the same as it would be if we performed a crossing change.

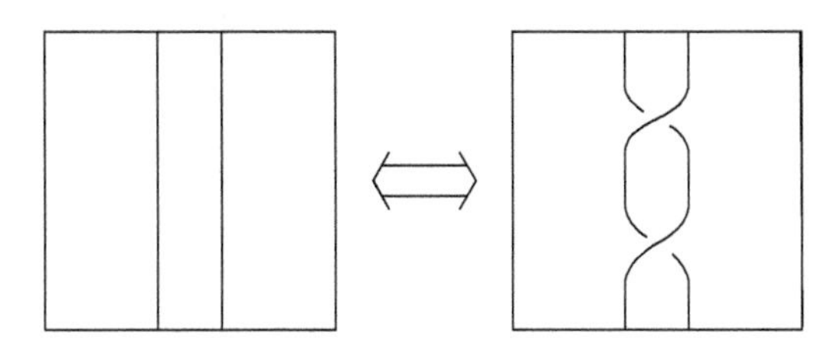

Figure 4.31 A local move on 1-links.

This is no accident. Note that the left-hand image of Figure 4.31 is the same as the the left-hand image of Figure 4.32.

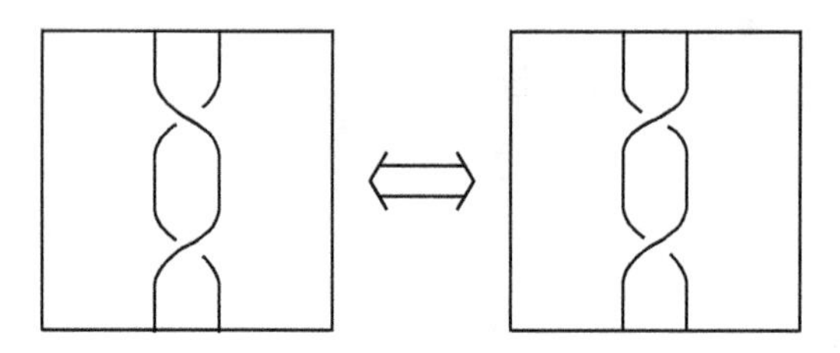

Figure 4.32 Crossing change.

See in Figure 4.33.

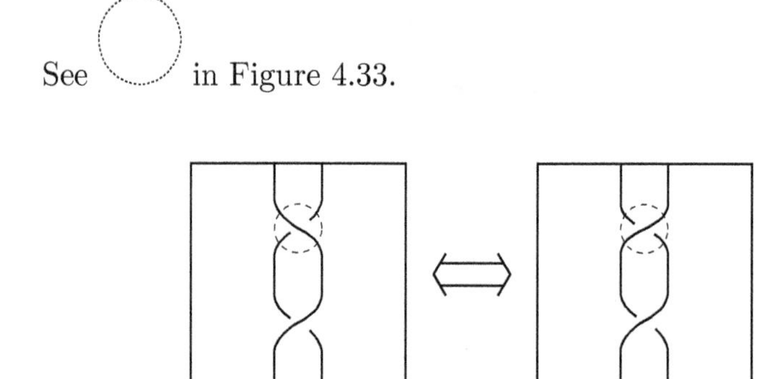

Figure 4.33 Crossing change.

Take a band $B^1 \times B^1$, labeled as X on the left-hand side of Figure 4.34 and Y on the right-hand side.

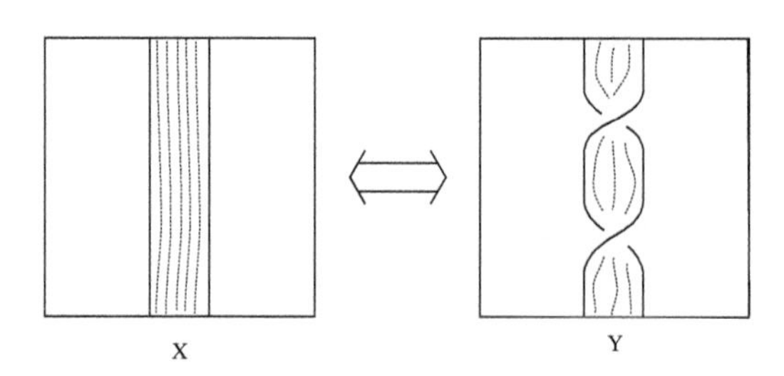

Figure 4.34 Bands, X and Y.

We will use this band to define the *twist-move* on $(2m + 1)$-dimensional knots in $\mathbb{R}^{2m+3}$.

Take a $(2m + 3)$-ball B^{2m+3} trivially embedded in $\mathbb{R}^{2m+3}$ and regard B^{2m+3} as $B^1 \times B^{m+1} \times B^{m+1}$. Also consider $B^{m+1} \times D^{m+1}$ trivially embedded in B^{2m+3} as shown in Figure 4.35. In this case, D^{m+1} (which is defined in page 61) is a smaller ball than B^{m+1}. Moreover, D^{m+1} is embedded in B^{m+1}. In Figure 4.35, this is drawn conceptually, albeit abstractly. We call X the embedding shown of $B^{m+1} \times D^{m+1}$.

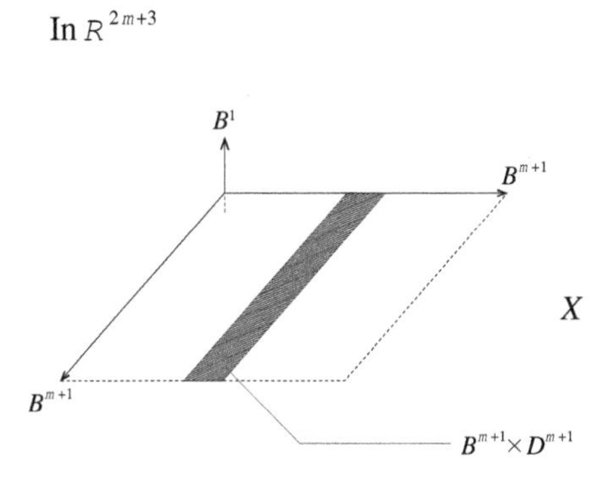

Figure 4.35 The twist-move. This figure continues in Figure 4.36.

Now, consider another embedding of $B^{m+1} \times D^{m+1}$ in B^{2m+3} as shown in Figure 4.36.

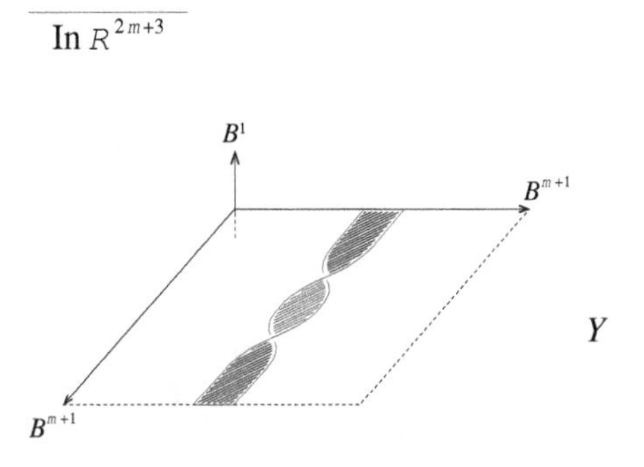

Figure 4.36 The twist-move. This figure is a continuation of Figure 4.35.

Note that this second embedding shown above looks like the band Y from Figure 4.34!

In Figures 4.37 and 4.38, we draw a more concrete visualization of the twist-move. We do not draw shaded part there.

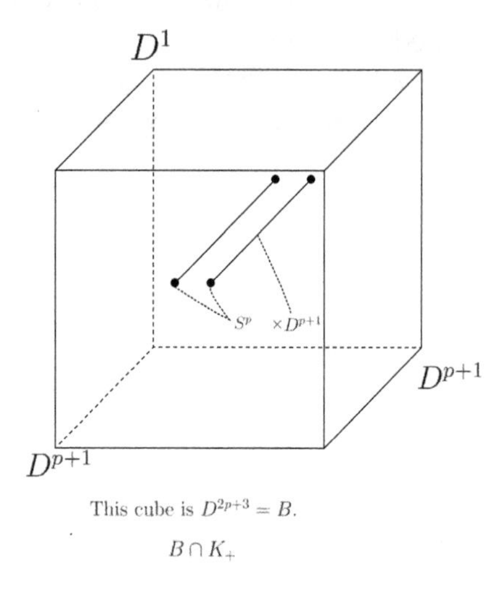

Figure 4.37 The twist-move. This figure continues in Figure 4.38.

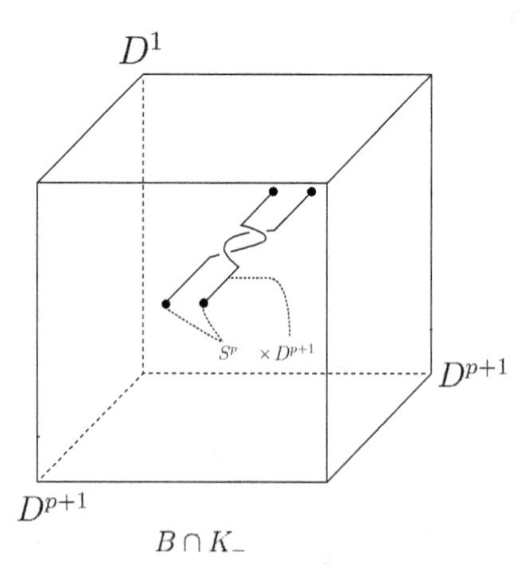

Figure 4.38 The twist-move. This figure is a continuation of Figure 4.37.

So, we can "twist" the embedding of $B^{m+1} \times D^{m+1}$ (the shaded part) in Figure 4.35 one full time around to obtain the embedding (the shaded part) shown in Figure 4.36. In the literature, we use something called a *Seifert pairing* to explicate the twist-move. See the author's paper [87] for further reading on this notion.

For now, consider the $2m+1 =$ one-dimensional case as shown in Figures 4.30–4.34. We want to use the one-dimensional case to help imagine the higher-dimensional case.

Note that the boundary of D^{m+1} is an m-sphere S^m. Therefore, $B^{m+1} \times S^m$ is included in the boundary of $B^{m+1} \times D^{m+1}$. We thus have $B^{m+1} \times S^m$ in the boundaries of our bands X and Y. Call X' that which is included in the boundary of X, and Y' that which is included in the boundary of Y.

Let K_+ and K_- be $(2m+1)$-dimensional knots in $\mathbb{R}^{2m+3}$. Suppose that K_+ and K_- differ only in B^{2m+3} and that the intersection of K_+ (respectively, K_-) and B^{2m+3} is X' (respectively, Y'). Then we say formally that K_+ (respectively, K_-) is obtained from K_- (respectively, K_+) by one twist-move. However, K_- is *not*, in general, the standard $(2m+1)$-sphere. (It may be an exotic sphere. See [74] for exotic spheres.) For more on this, see the author's paper [87].

Note: There exist infinitely many non-trivial n-knots which are twist-move equivalent to the trivial n-knot — i.e., n-knots which can be transformed into the trivial n-knot by a finite sequence of twist-moves. On the other hand, there also exist infinitely many non-trivial n-knots which are *not* twist-move equivalent to the trivial n-knot.

In Chapter 1, we discussed some invariants and polynomials of 1-links which have a tight relationship with the local moves we described.

So, you may wonder: Do we have a relationship between local moves on n-links and invariants and polynomials on n-links?

Absolutely! We discuss these relationships in the following sections. One of them is the author's result [44], a new type of local move identity for the Alexander polynomial of high-dimensional knots. The identity is as follows:

$$\mathrm{Alex}(K_+) - \mathrm{Alex}(K_-) = (t+1) \cdot \mathrm{Alex}(K_0).$$

Note: We use $(t+1)$ instead of $(t-1)$, as is perhaps expected. This identity is associated with the twist-move.

We have some other relations as well; we go through them in the subsequent chapters.

Chapter 5

Local Moves on High-Dimensional Knots and Related Invariants

5.1 Spun Knots and Twist-Spun Knots

As we stated in Section 4.1 of Chapter 4, there is a 2-knot that is not ribbon-move equivalent to the trivial 2-knot. In this section, we will show such an example.

We show a rotation which produces four-dimensional space $\mathbb{R}^4$ before we introduce spun and twist-spun knots

5.1.1 *A rotation makes* $\mathbb{R}^4$

Here, we show a method to obtain $\mathbb{R}^4$ by a kind of rotation.

Note that we can regard $\mathbb{R}^2$ as the result of rotating $\mathbb{R}_{\geq 0} = \{(x, y)|x \geqq 0,\ y = 0\}$ around the point $(0, 0)$. See Figure 5.1.

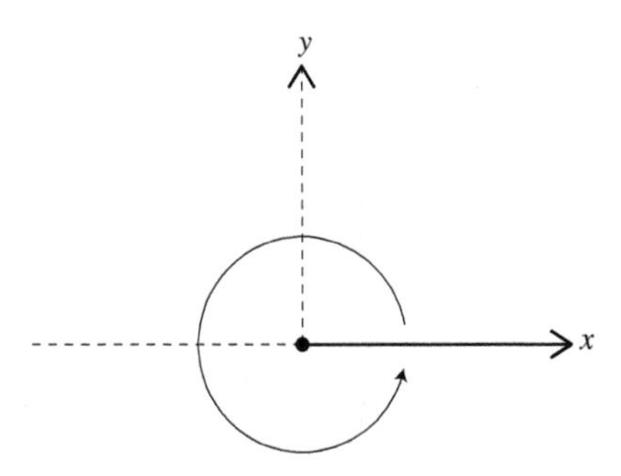

Figure 5.1 $\mathbb{R}^2$ is the result of rotating $\mathbb{R}^1_{\geq 0} = \{(x, y)|x \geqq 0,\ y = 0\}$ around the point $(0, 0)$.

Similarly, we can regard $\mathbb{R}^3$ as the result of rotating $\mathbb{R}^2_{\geq 0} = \{(x, y, z)|x \geqq 0,\ y = 0\}$ around the z-axis. See Figure 5.2. We draw the x-, y-, and z-axes in a different fashion from what you are used to.

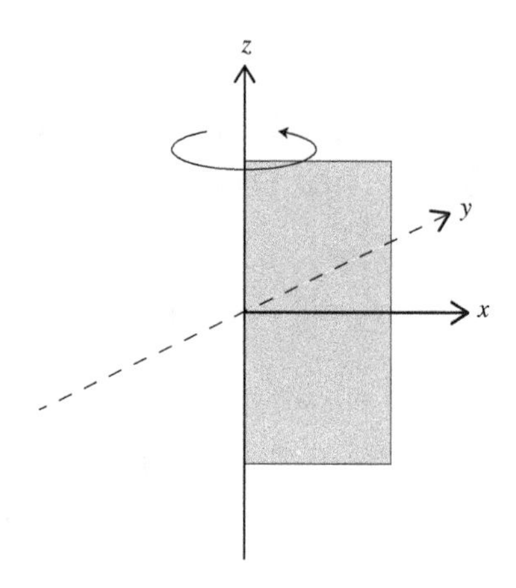

Figure 5.2 $\mathbb{R}^3$ is the result of rotating $\mathbb{R}^2_{\geq 0} = \{(x, y, z)|x \geqq 0,\ y = 0\}$ around the z-axis.

Now consider Figure 5.3. Rotate the arc around the dotted line in $\mathbb{R}^3$ — the result is the 2-sphere S^2 in $\mathbb{R}^3$.

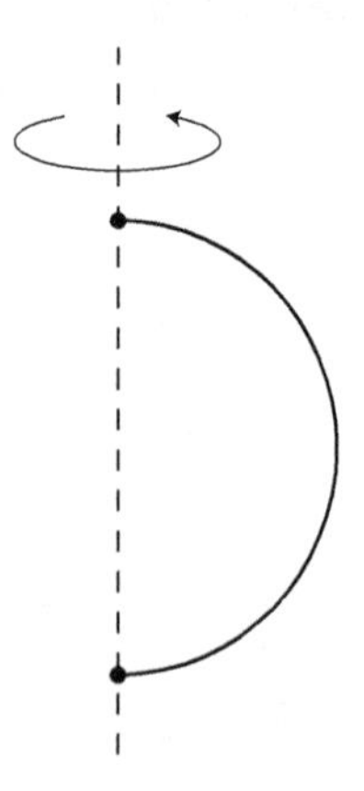

Figure 5.3 If we rotate the arc around the dotted line in $\mathbb{R}^3$, then the result is S^2.

Suppose that there is a point in

$$\mathbb{R}^2_{\geq 0} = \{(x, y, z)|x \geqq 0,\ y = 0\}$$

as shown in Figure 5.4. When we rotate $\mathbb{R}^2_{\geq 0}$ around the z-axis, rotate the point as well. The result is S^1 in $\mathbb{R}^3$.

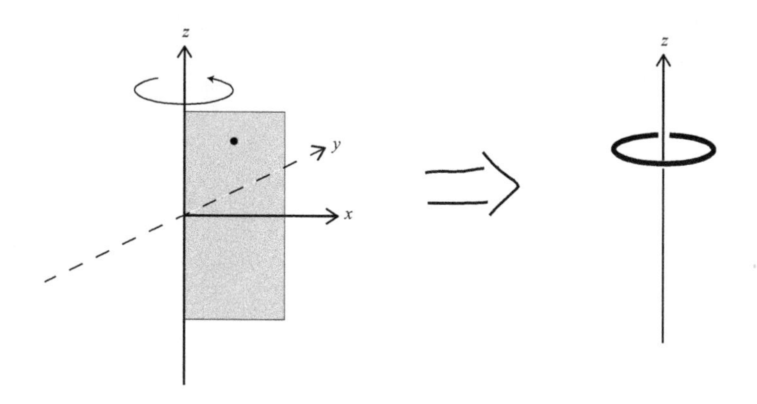

Figure 5.4 A circle is made by the rotation of a point around the z-axis.

Now, we think of a new way to visualize $\mathbb{R}^4$: rotate $\mathbb{R}^3_{\geq 0} = \{(x, y, z, w)|x \geqq 0,\ y = 0\}$ around the zw-plane. See Figure 5.5. Try to imagine this in action!

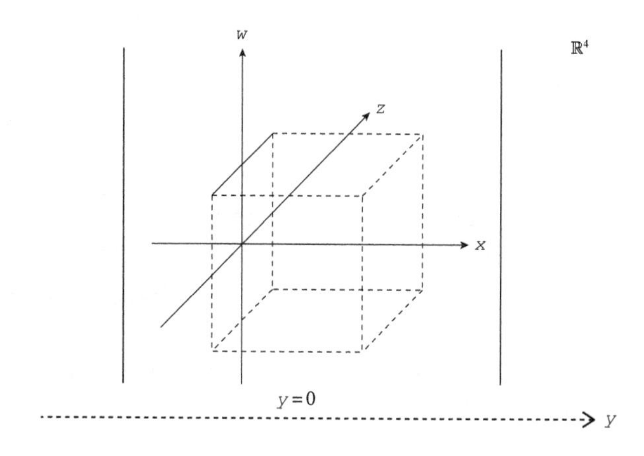

Figure 5.5 We regard $\mathbb{R}^4$ as the result of rotating $\mathbb{R}^3_{\geq 0} = \{(x, y, z, w)|x \geqq 0,$ $y = 0\}$ around the zw-plane.

5.1.2 *Defining two-dimensional spun knots and twist-spun knots*

Suppose that there is an arc in $\mathbb{R}^3_{\geq 0} = \{(x, y, z, w) | x \geqq 0,\ y = 0\}$ as shown in Figure 5.6.

When we rotate $\mathbb{R}^3_{\geq 0}$ around the zw-plane, rotate the arc as well. We obtain S^2 in $\mathbb{R}^4$. Try to imagine this! More rather, we have obtained a 2-knot.

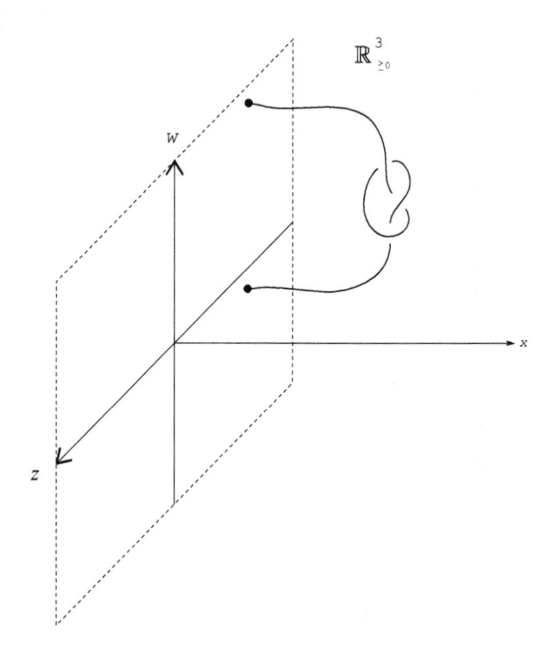

Figure 5.6 A rotation to make a 2-knot in $\mathbb{R}^4$.

Let A and B be the endpoints of the arc. Suppose that A and B are in the wz-plane. Connect A with B by the dotted segment as shown in Figure 5.7. From the arc and the dotted segment we get a 1-knot — call it K.

The 2-knot produced from this arc as shown in Figure 5.7 is called the *spun knot* of the 1-knot K.

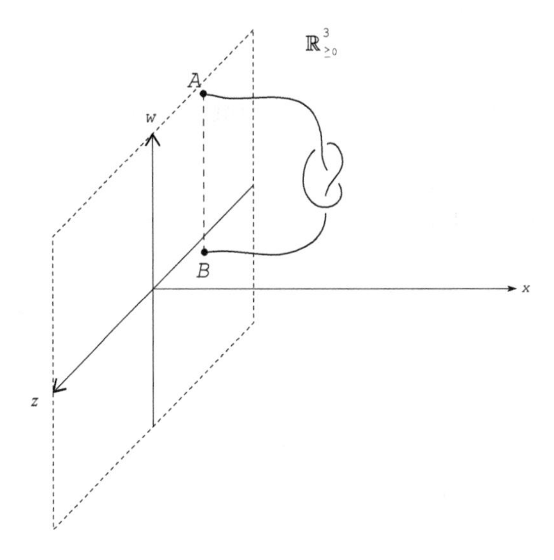

Figure 5.7 Creating a spun knot.

It is known that if K is a non-trivial 1-knot, then the spun knot of K is a non-trivial 2-knot.

It is also known that there are countably infinitely many different 1-knots whose spun knots are different from one another. Hence, there are countably infinitely many different 2-knots.

See [61, 117, 118, 131] for more details on this topic.

See Figure 5.8. When we rotate $\mathbb{R}^3_{\geq 0}$ as well as the given arc around the zw-plane, rotate the part of the arc between P and Q k-times as shown in Figure 5.8, for some $k \in \mathbb{Z}$. Note that we are rotating the 3-ball that is represented by the dotted sphere in Figure 5.8, and that includes the knotted part of the PQ-segment of the arc. We can think of this 3-ball as a planet in our solar system; this operation is like the planet's revolution around the sun and simultaneous rotation about its own central axis.

From the process shown in Figure 5.8, we obtain a 2-knot in $\mathbb{R}^4$. The resulting knot is called the *k-twist-spun knot* of the 1-knot K. Note that ordinary spun knots as defined previously are actually just 0-twist-spun knots.

Zeeman [131] proved that the k-twist-spun knot of any 1-knot is the trivial 2-knot ($k = \pm 1$).

It is known that there are countably infinitely many different 2-knots that are k-twist-spun knots if $k \neq 0, \pm 1$.

Furthermore, it is known that there are countably infinitely many non-trivial 2-knots which are not a k-twist-spun knot for any integer k.

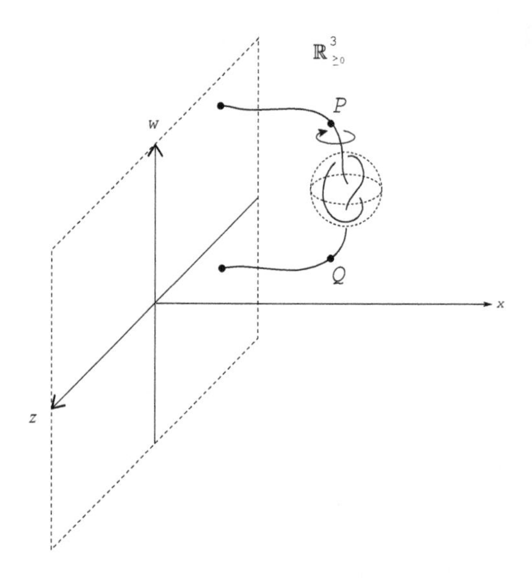

Figure 5.8 Making a twist-spun knot from this.

See the following sections. See [61, 117, 118, 131] for further reading.

5.1.3 *Defining high-dimensional spun knots and twist-spun knots*

Refer to Figure 5.9. We regard $\mathbb{R}^{n+2}$ as the result of rotating

$$\mathbb{R}^{n+1}_{\geq 0} = \{(x_1, \ldots, x_{n+2}) | x_1 \geqq 0,\ x_2 = 0\}$$

around the $x_3 \cdots x_{n+2}$-space

$$\{(x_1, \ldots, x_{n+2}) | x_1 = 0,\ x_2 = 0\},$$

which is the n-space $\mathbb{R}^n$.

Take an $(n-1)$-ball B^{n-1} in $\mathbb{R}^{n+1}_{\geq 0}$ and an $(n-1)$-ball B_0^{n-1} in the $x_3 \cdots x_{n+2}$-space such that the boundary of B^{n-1} is that of B_0^{n-1} in the $x_3 \cdots x_{n+2}$-space. We obtain an $(n-1)$-knot K from B^{n-1} and B_0^{n-1}.

See Figure 5.9. When we rotate $\mathbb{R}^{n+1}_{\geq 0}$ around the $x_3 \cdots x_{n+2}$-space, rotate B^{n-1} as well. The result is S^n in $\mathbb{R}^{n+2}$. Try to picture this! Hence, we obtain an n-knot from the $(n-1)$-knot K. This n-knot is called the *spun knot* of K. This is merely a generalization of our low-dimensional definition of spun knots to higher-dimensional space.

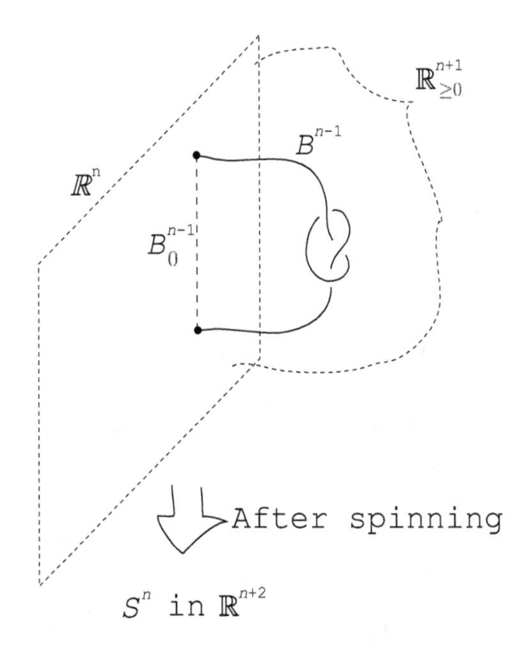

Figure 5.9 An n-dimensional spun knot.

It is known that there are countably infinitely many different $(n-1)$-knots whose spun knots are different from one another. Thus, there are countably infinitely many different n-knots ($n \geqq 3$). This result is analogous to one we described for the low-dimensional case.

See [61, 117, 118, 131] for further reading on this matter.

For any higher-dimensional knot K, we can define the *k-twist spun knot* of K as follows. See Figure 5.10 to help visualize the higher-dimensional case. When we rotate B^{n-1} around the $x_3 \cdots x_{n+2}$-space, we rotate a part of B^{n-1} in a similar fashion to the case shown in Figure 5.8, illustrating how to make k-twist-spun knots from 1-knots.

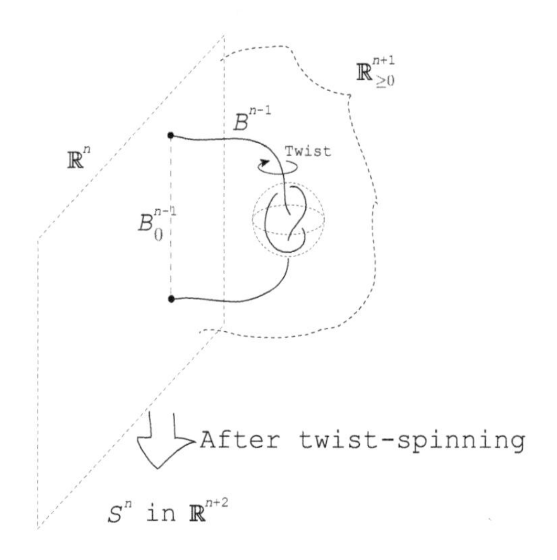

Figure 5.10 An n-dimensional twist-spun knot.

Zeeman [131] proved that the ± 1-twist-spun knot of any $(n-1)$-knot is the trivial n-knot.

It is known that there are countably infinitely many different n-knots ($n \geqq 3$) which are k-twist-spun knots for each $k \neq 0, \pm 1$.

Furthermore, it is known that there are countably infinitely many non-trivial n-knots ($n \geqq 3$) which are not k-twist-spun knots for any integer k.

See [61, 117, 118, 131] for more details.

5.2 The μ- and $\mathbb{Q}/\mathbb{Z}$-$\widetilde{\eta}$-Invariants

For any given 2-knot, we can define the μ-invariant and the $\mathbb{Q}/\mathbb{Z}$-$\widetilde{\eta}$-invariants. See Refs. [85, 90, 118]. The ribbon-moves of 2-knots in Section 4.1 of Chapter 4 preserve both of these types of invariants for 2-knots in $\mathbb{R}^4$.

Many two-dimensional twist-spun knots of 1-knots have non-trivial μ-invariant or non-trivial $\mathbb{Q}/\mathbb{Z}$-$\widetilde{\eta}$-invariants. For example, the 5-twist-spun knot of the trefoil knot has non-trivial μ-invariant. Moreover, those 2-knots are not ribbon-move equivalent to the trivial knot.

5.3 Ribbon-Move Unknotting Number of 2-Knots in $\mathbb{R}^4$

Let K be a 2-knot in $\mathbb{R}^4$. Assume that n ribbon-moves unknot K, but that no sequence of p ribbon-moves unknot K (for $p < n$). Then we say that the *ribbon (unknotting) number* of K is n.

As we state in Section 4.1 of Chapter 4, there is a 2-knot whose ribbon number is two. We get an example from taking a connected sum of two spun knots of the trefoil knot. This fact was proved by the author [97].

The author [97] also proved that the ribbon-move unknotting number is unbounded.

Formally, we define a *connected sum* of two 1-knots as follows. Let J and K be 1-knots in $\mathbb{R}^3$, whose coordinates are given by t, x, y. Assume that J is in $t < 0$ and that K is in $t > 0$. The disjoint union of J and K is a 2-component 1-link $L = (J, K)$ in $\mathbb{R}^3$. See the upper image of Figure 5.11.

Refer to Figure 5.11. We want to change the upper illustration into the lower one. In doing so, we obtain a new 1-knot from J and K. This new 1-knot is the *connected sum* $J\sharp K$ of two 1-knots J and K.

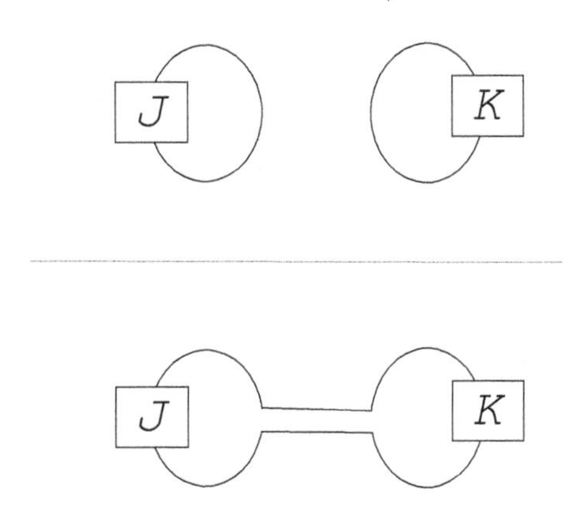

Figure 5.11 The connected sum of two 1-knots.

Formally, we define a *connected sum* of two 2-knots as follows. Let J and K be 2-knots in $\mathbb{R}^4$, whose coordinates are given by t, x, y, z. Assume that J is in $t < 0$ and that K is in $t > 0$. The union of J and K is a 2-component 2-link $L = (J, K)$ in $\mathbb{R}^4$. Take a 4-ball B in $\mathbb{R}^4$ such that $B \cap J$ and $B \cap K$ appear as shown in the upper image of Figure 5.12: They are two discs. One disc is a part of J, and the other, a part of K.

Refer to Figure 5.12. We want to change the upper illustration into the lower one. Remove a small disc from each disc and connect by using a tube. The result is one annulus or one tube since we can transform this object smoothly without touching it, into the annulus. In doing so, we obtain a new 2-knot from J and K. This new 2-knot is the *connected sum* $J \sharp K$ of two 2-knots J and K.

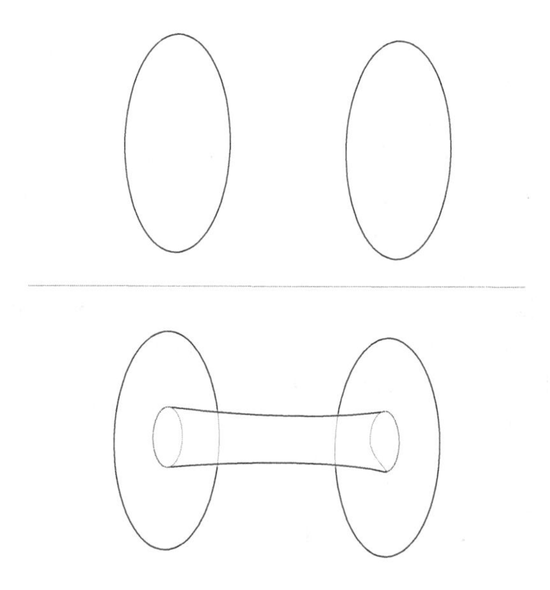

Figure 5.12 The connected sum.

5.4 The Signature and the Arf Invariant

If $p = q$, the (p, q)-pass-move on a $(p + q - 1)$-knot in $\mathbb{R}^{p+q+1}$ is called the (*high-dimensional*) *pass-move* on an n-knot in $\mathbb{R}^{n+2}$, where $n = p + q - 1$.

The pass-move on $(4k+1)$-knots in $\mathbb{R}^{4k+3}$ preserves what we call the *Arf invariant* of $(4k+1)$-knots. Respectively, the pass-move on $(4k+3)$-knots in $\mathbb{R}^{4k+5}$ preserves what we call the *signature* of $(4k+3)$-knots. There are infinitely many $(4k+1)$-knots and $(4k+3)$-knots whose Arf invariants and signatures, respectively, are non-trivial. These knots are not unknotted by a sequence of pass-moves. See the author's papers [79, 97].

As we wrote in Section 4.3 of Chapter 4, the author gives a necessary and sufficient condition that $(4k+1)$-knots in $\mathbb{R}^{4k+3}$ and $(4k+3)$-knots in $\mathbb{R}^{4k+5}$ are pass-move equivalent to the trivial knot in their respective spaces. See [Ref. [44], Proposition 12.1] and [Ref. [79], Theorem 4.1].

In Section 1.4 of Chapter 1, we described a relation between local moves on 1-knots and polynomials of 1-knots. We elucidate this concept in Chapter 6 before we introduce the relationships between local moves and polynomials for high-dimensional knots in Chapter 7.

Chapter 6

The Alexander and Jones Polynomials of One-Dimensional Links in $\mathbb{R}^3$

6.1 Oriented Knots and Links

To each knot, we give an *orientation*, denoted by an arrow. This arrow indicates the direction in which we travel along the knot. There are two ways to give an orientation to each knot.

Recall that in Section 1.3 of Chapter 1, we use orientations of knots and links when we introduce pass-moves.

For a knot K, $-K$ is the knot with the opposite orientation to that of K. We can define *isotopy* of oriented links as we define that of (non-oriented) links in page 15.

There actually do exist knots J such that J and $-J$ are isotopic. The trivial knot is such an example. See Figure 6.1. We transform one to the other by an isotopy; this is drawn in Figure 6.2.

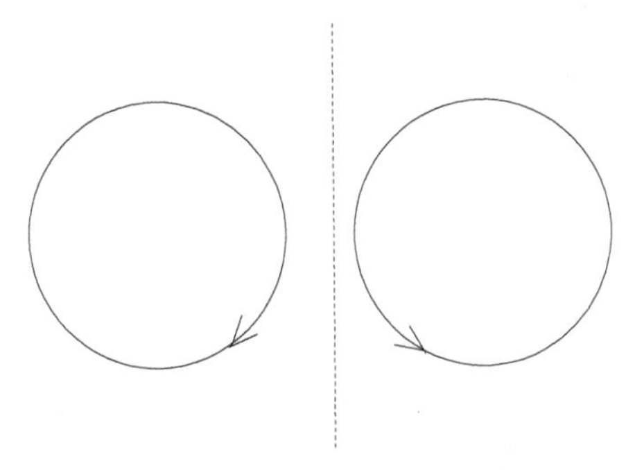

Figure 6.1 Endowing the trivial knot with opposite orientations.

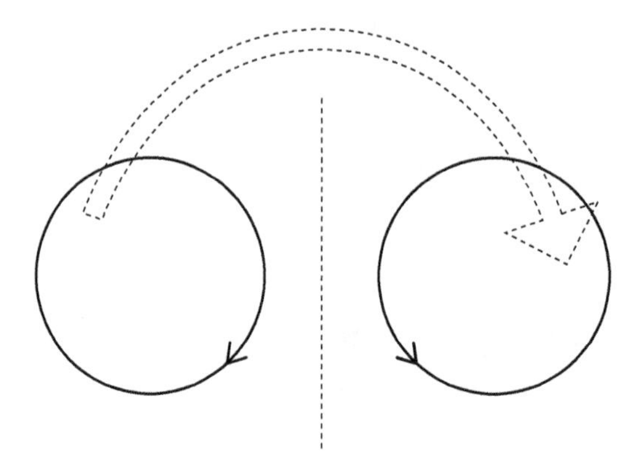

Figure 6.2 An isotopy between oppositely oriented knots.

The trivial knot is not the only knot with this property. Let K be the trefoil knot endowed with an orientation, and let $-K$ be the trefoil knot with the opposite orientation. See Figure 6.3. K and $-K$ are isotopic. We transform one to the other by an isotopy, drawn in Figure 6.4.

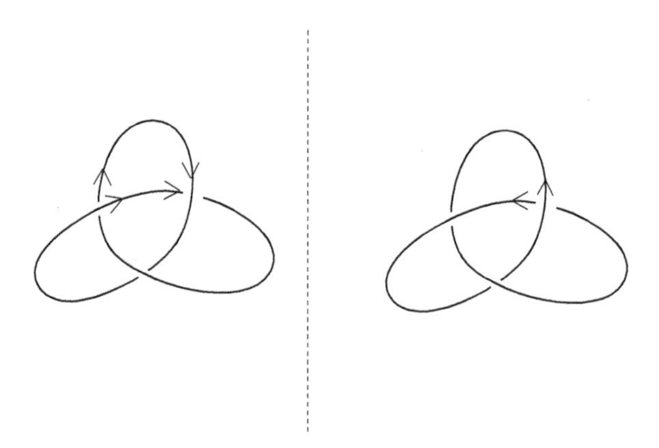

Figure 6.3 The trefoil knot with two opposite orientations.

However, this property is not always true. Trotter [126] proved that there does exist a knot K such that K and $-K$ are not isotopic.

As for an n-component link, each component has two possible orientations. Hence, there are 2^n ways to orient our given link.

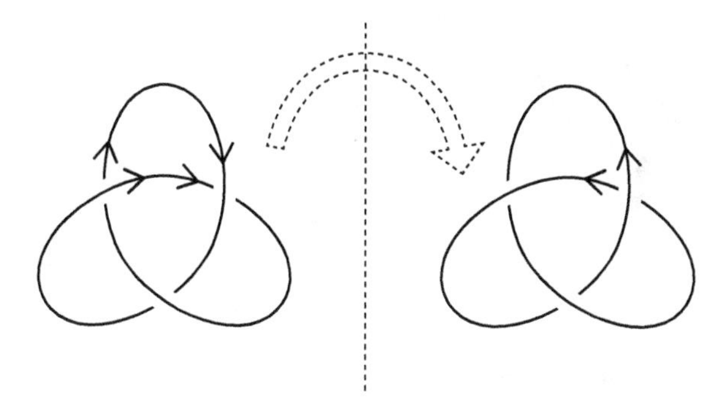

Figure 6.4 An isotopy between oppositely oriented trefoil knots.

For example, there are four ways to orient the trivial 2-component link. One of four is drawn in Figure 6.5. The four oriented links are all isotopic. The proof follows the same line of reasoning as that illustrated in Figures 6.2 and 6.4.

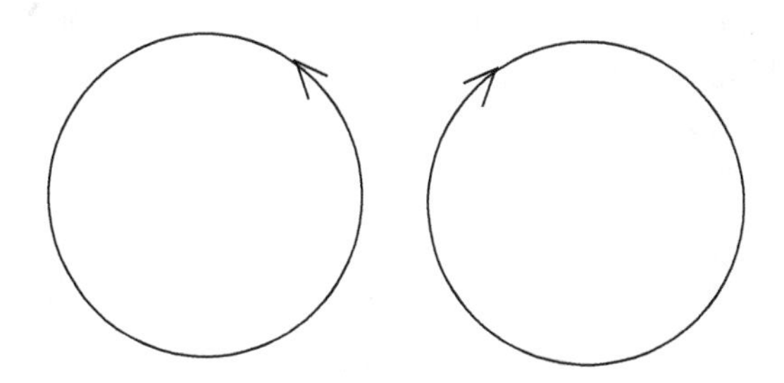

Figure 6.5 The oriented trivial 2-component link.

You may ask: Are the following two knots in Figure 6.6 isotopic?

Remark: Each knot does not change by switching the orientation.

Note the difference between the crossing points of the knots shown in Figure 6.6. We will answer this question in Section 6.6. (In this case, the difference in crossings turns out to be fundamental to the knot diagram!)

Figure 6.6 Are these two knots the same?

6.2 Defining the Alexander and Jones Polynomials in $\mathbb{R}^3$

We define the Alexander and Jones polynomials for knots and links in $\mathbb{R}^3$.

Definition 6.1. Let K_+, K_-, and K_0 be oriented links in $\mathbb{R}^3$. The orientation of each component is indicated by an arrow.

Assume that the three links differ only in the 3-ball as drawn in Figure 6.7. This relation is called the *skein relation.*

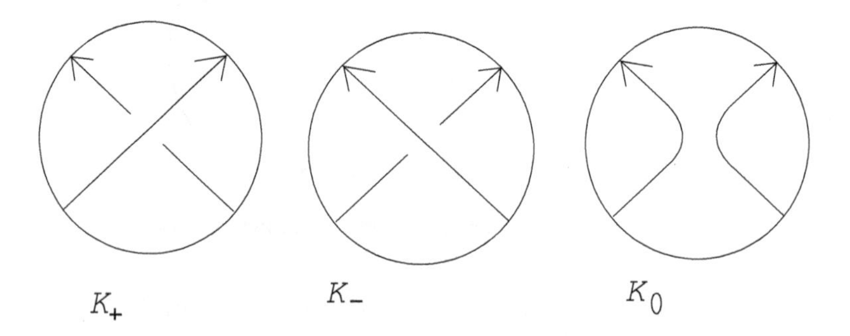

Figure 6.7 The skein relation.

To each knot K, we assign the *Alexander polynomial*, $\mathrm{Alex}(K)$, as well as the *Jones polynomial*, $V(K)$, uniquely satisfying the following properties:

(1) For the trivial knot T, $\text{Alex}(T) = V(T) = 1$.
(2) The following identities are satisfied by their respective polynomials:

$$\text{Alexander: } \text{Alex}_{K_+}(t) - \text{Alex}_{K_-}(t) = (t^{\frac{1}{2}} - t^{-\frac{1}{2}}) \cdot \text{Alex}_{K_0}(t);$$

$$\text{Jones: } t^{-1}V_{K_+}(t) - tV_{K_-}(t) = (t^{\frac{1}{2}} - t^{-\frac{1}{2}})V_{K_0}(t).$$

Note that Definition 6.1 does *not* claim that the Alexander or Jones polynomial is 1 for the trivial n-component link ($n > 1$).

We can prove the following: Every Alexander polynomial and every Jones polynomial is composed of a finite number of terms. And moreover, the degree $*$ of each term t^* of any Alexander or Jones polynomial is an integer or a half integer.

A set of three links in the skein relation is drawn in Figure 6.8.

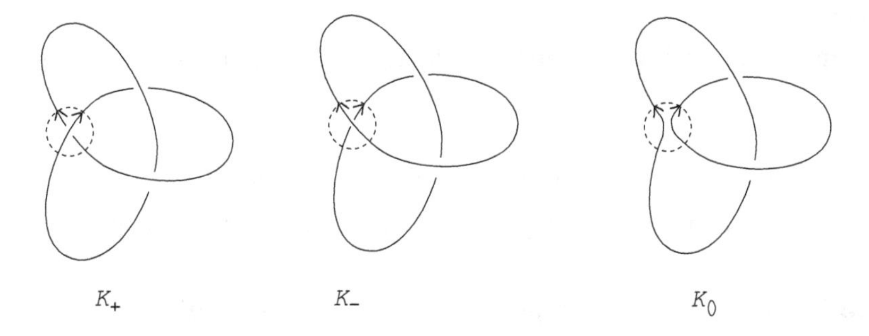

Figure 6.8 Three oriented link in the skein relation.

We have the following two theorems about the Alexander and Jones polynomials.

Theorem 6.2 (Alexander [1] and Jones [32]). *Definition 6.1 is well-defined. That is, by using Definition 6.1, we can assign to each oriented knot a unique polynomial. Moreover, we can calculate the polynomial explicitly.*

Kauffman [38] gave a very short alternative proof that the Jones polynomial is an isotopy-type invariant of 1-links in the 3-sphere.

We have thus established well-definedness. But better yet, we have the following:

Theorem 6.3 (Alexander [1] and Jones [32]).

(1) *Let J and K be oriented links. If J and K are isotopic, we have* $\mathrm{Alex}_{\mathrm{J}}(\mathrm{t}) = \mathrm{Alex}_{\mathrm{K}}(\mathrm{t})$ *(respectively, $V_J(t) = V_K(t)$).*
(2) *The Alexander (respectively, Jones) polynomial can distinguish infinitely many knots.*

Let K be an oriented knot. Make $-K$ from K. It is a known fact that the Alexander and Jones polynomials of knots, or one-component links, are invariant under switching the orientation. More rather, K and $-K$ have the same Alexander polynomial and the same Jones polynomial. The proof is quite accessible. Consider how the crossings shown in Figure 6.7 change when we reverse the orientation.

However, in the case of links with more than one component, the associated polynomials may in fact depend on the orientation. See Section 6.5.

In the following sections, we go through explicit calculations of the Alexander and Jones polynomials.

6.3 Calculating the Alexander Polynomial

We calculate the Alexander polynomials of some familiar links.

6.3.1 *The Alexander polynomial of the trivial 2-component link*

Orient the trivial 2-component link as shown in Figure 6.9. Let us now use Definition 6.1 to calculate the Alexander polynomial.

Recall: Definition 6.1(1) claims that the Alexander polynomial of the trivial knot — that is, the trivial one-component link — is 1. However, it does *not* claim that the Alexander polynomial of the trivial 2-component link is 1.

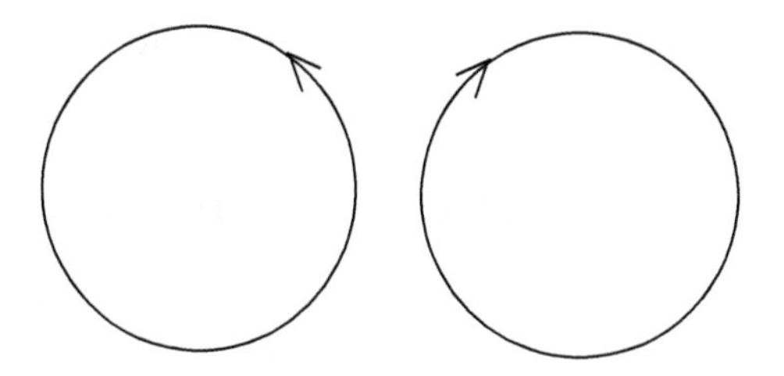

Figure 6.9 The oriented trivial 2-component link.

See Figure 6.10. We have three oriented links, K_+, K_-, and K_0 in the skein relation. By Definition 6.1, the Alexander polynomials

$$\Delta_{K_+}(t), \Delta_{K_-}(t), \Delta_{K_0}(t)$$

satisfy the relation

$$\Delta_{K_+}(t) - \Delta_{K_-}(t) = (t^{\frac{1}{2}} - t^{-\frac{1}{2}}) \cdot \Delta_{K_0}(t).$$

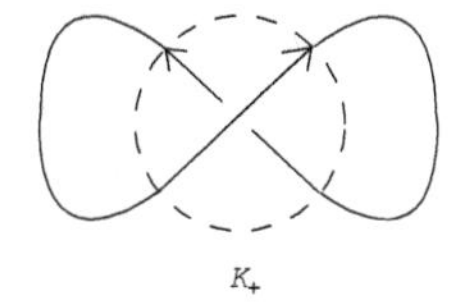

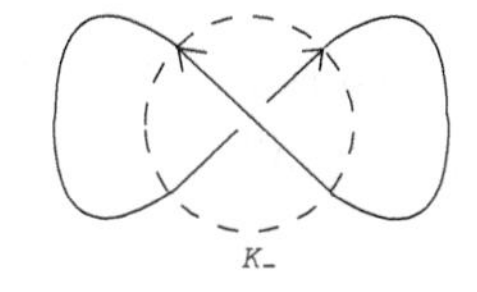

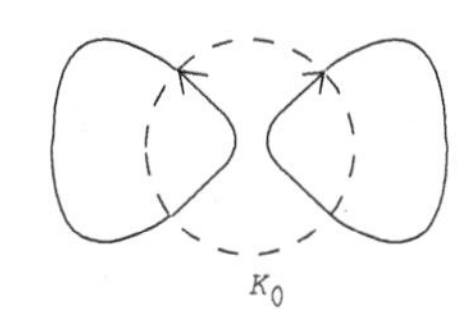

Figure 6.10 Three oriented links, K_+, K_-, and K_0.

Both K_+ and K_- are the trivial knot, that is, the trivial one-component link. Hence, we have

$$\Delta_{K_+}(t) = 1, \quad \Delta_{K_-}(t) = 1.$$

K_0 is the trivial 2-component link oriented as drawn in Figure 6.9. We calculate the Alexander polynomial $\Delta_{K_0}(t)$ of K_0 as follows:

$$\begin{aligned}(t^{\frac{1}{2}} - t^{-\frac{1}{2}})\Delta_{K_0}(t) &= \Delta_{K_+}(t) - \Delta_{K_-}(t) \\ &= 1 - 1 \\ &= 0.\end{aligned}$$

Therefore,

$$\Delta_{K_0}(t) = 0.$$

Thus, the Alexander polynomial of the trivial 2-component link oriented as drawn in Figure 6.9 is 0.

Recall that there are four ways to orient the trivial 2-component link, and that the four oriented links are all isotopic. See page 85.

6.3.2 *The Alexander polynomial of the Hopf link*

We calculate the Alexander polynomial of the Hopf link, endowing it with the orientation shown in Figure 6.11.

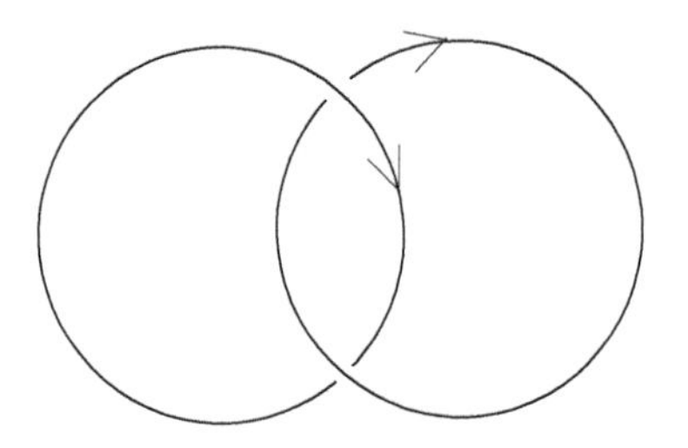

Figure 6.11 The Hopf link.

See Figure 6.12. Three oriented links, K_+, K_-, and K_0, are in the skein relation. We again recall Definition 6.1 to see that the Alexander polynomials

$$\Delta_{K_+}(t), \Delta_{K_-}(t), \Delta_{K_0}(t)$$

satisfy the relation

$$\Delta_{K_+}(t) - \Delta_{K_-}(t) = (t^{\frac{1}{2}} - t^{-\frac{1}{2}}) \cdot \Delta_{K_0}(t).$$

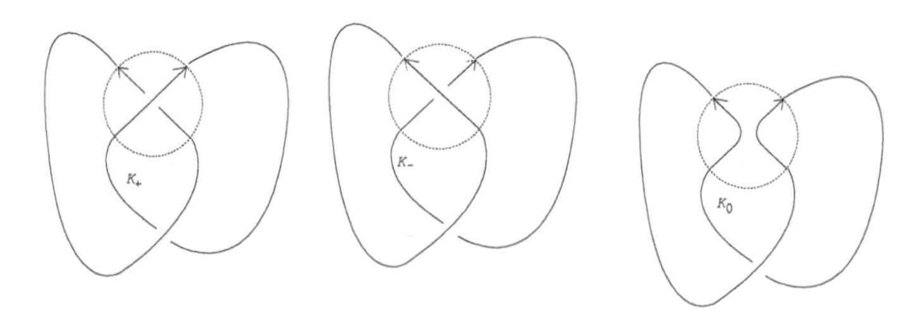

Figure 6.12 Three oriented links, K_+, K_-, and K_0.

K_- is isotopic to the trivial 2-component link oriented as drawn in Section 6.3.1. The Alexander polynomial $\Delta_{K_-}(t)$ is 0 as we calculated in Section 6.3.1. K_0 is the trivial knot. By Definition 6.1(1), the Alexander polynomial $\Delta_{K_0}(t)$ is 1.

Lastly, we have K_+, the Hopf link shown in Figure 6.12. We calculate the Alexander polynomial $\Delta_{K_+}(t)$:

$$\begin{aligned}\Delta_{K_+}(t) &= \Delta_{K_-}(t) + (t^{\frac{1}{2}} - t^{-\frac{1}{2}})\Delta_{K_0}(t)\\ &= 0 + (t^{\frac{1}{2}} - t^{-\frac{1}{2}}) \cdot 1\\ &= t^{\frac{1}{2}} - t^{-\frac{1}{2}}.\end{aligned}$$

6.3.3 *The Alexander polynomial of the trefoil knot*

We calculate the Alexander polynomial of the trefoil knot, oriented as drawn in Figure 6.13.

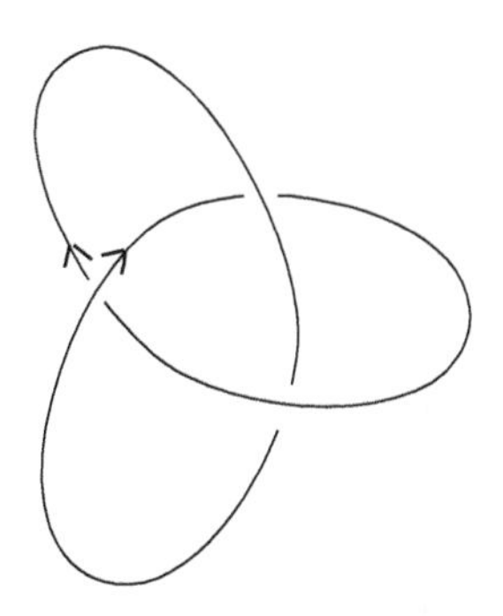

Figure 6.13 The trefoil knot.

See Figure 6.14.

Three oriented links, K_+, K_-, and K_0, are in the skein relation.

Again, by Definition 6.1, we get that the Alexander polynomials

$$\Delta_{K_+}(t), \Delta_{K_-}(t), \Delta_{K_0}(t)$$

satisfy the relation

$$\Delta_{K_+}(t) - \Delta_{K_-}(t) = (t^{\frac{1}{2}} - t^{-\frac{1}{2}}) \cdot \Delta_{K_0}(t).$$

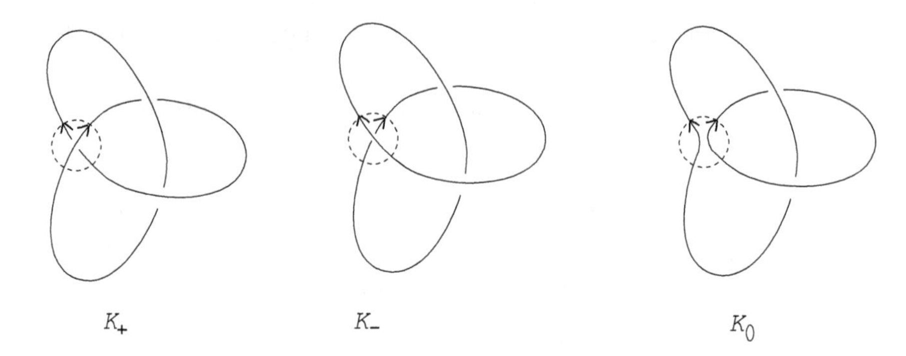

Figure 6.14 Three oriented links, K_+, K_-, and K_0.

See Figure 6.14. K_- is the trivial knot. By Definition 6.1(1), the Alexander polynomial $\Delta_{K_-}(t)$ of K_- is 1. K_0 is isotopic to the Hopf link oriented as drawn in Figure 6.11. As we calculated in Section 6.3.2, the Alexander polynomial $\Delta_{K_0}(t)$ of K_0 is $t^{\frac{1}{2}} - t^{-\frac{1}{2}}$.

K_+ is the trefoil knot oriented as drawn in Figure 6.13. We thus calculate the Alexander polynomial $\Delta_{K_+}(t)$ of K_+ as follows:

$$\begin{aligned}\Delta_{K_+}(t) &= \Delta_{K_-}(t) + (t^{\frac{1}{2}} - t^{-\frac{1}{2}})\Delta_{K_0}(t)\\ &= 1 + (t^{\frac{1}{2}} - t^{-\frac{1}{2}})(t^{\frac{1}{2}} - t^{-\frac{1}{2}})\\ &= t - 1 + t^{-1}.\end{aligned}$$

Recall that it is not always possible to calculate the unknotting number (page 19) of a given knot. However, we *can* calculate the Alexander polynomial for all links by the same method as we have used for the trivial link, the Hopf link, and the trefoil knot.

Remark: If the Alexander polynomials of two knots are different, we can conclude that the two knots are not isotopic.

Calculate the Alexander polynomial of the knot shown in Figure 6.15. This knot is called the figure-eight knot.

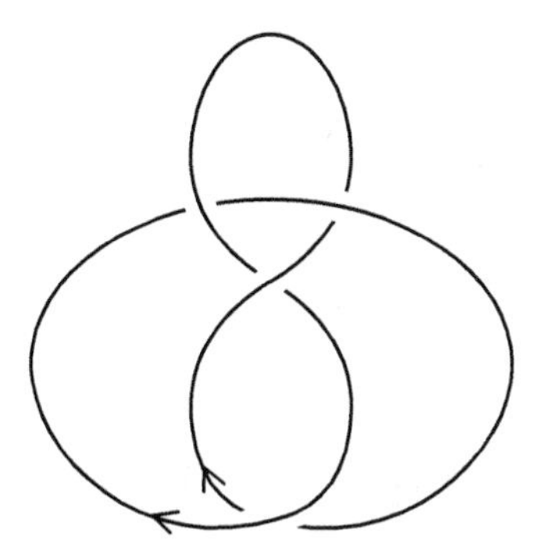

Figure 6.15 The figure-eight knot.

6.4 Calculating the Jones Polynomial

We have now gone through some explicit calculations of the Alexander polynomial. In this section, we do the same for the Jones polynomial.

The method of calculating the Jones polynomial is similar to that for the Alexander polynomial, though it is important to keep in mind that there are infinitely many knots such that the Alexander polynomial and the Jones polynomial are different.

Furthermore, there exists a pair of knots J and K such that $\text{Alex}(J) = \text{Alex}(K)$ but $V(J) \neq V(K)$, and some pair J' and K' such that $V(J) = V(K)$ but $\text{Alex}(J) \neq \text{Alex}(K)$.

6.4.1 *The Jones polynomial of the trivial 2-component link*

We calculate the Jones polynomial of the trivial 2-component link, oriented as drawn in Figure 6.16.

Recall: Definition 6.1(1) claims that the Jones polynomial of the trivial knot — that is, the trivial one-component link — is 1, but not that the Jones polynomial of the trivial 2-component link is 1.

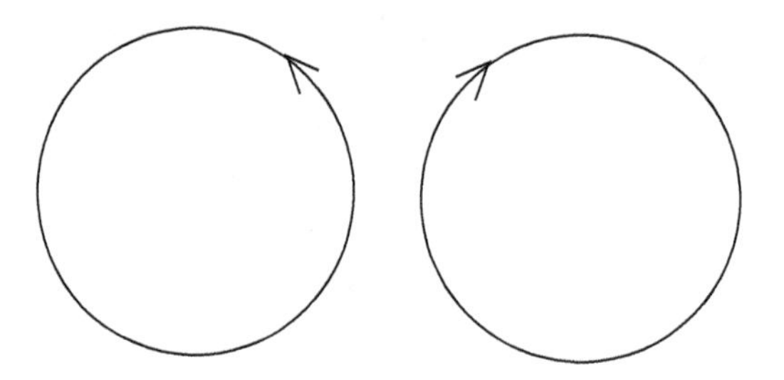

Figure 6.16 The oriented trivial 2-component link. This figure is the same as Figure 6.9.

Refer to Figure 6.17. Three oriented links, K_+, K_-, and K_0, are shown in the skein relation. Definition 6.1 tells us that the Jones polynomials

$$V_{K_+}(t), V_{K_-}(t), V_{K_0}(t)$$

satisfy the relation

$$t^{-1}V_{K_+}(t) - tV_{K_-}(t) = (t^{\frac{1}{2}} - t^{-\frac{1}{2}})V_{K_0}(t).$$

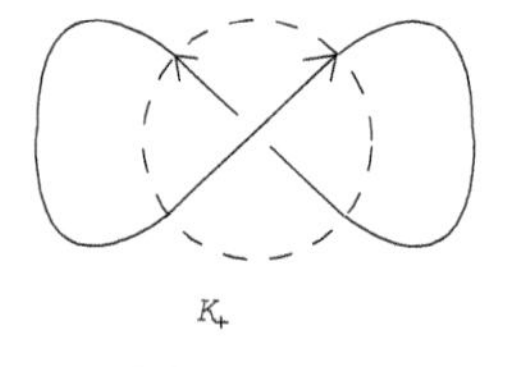

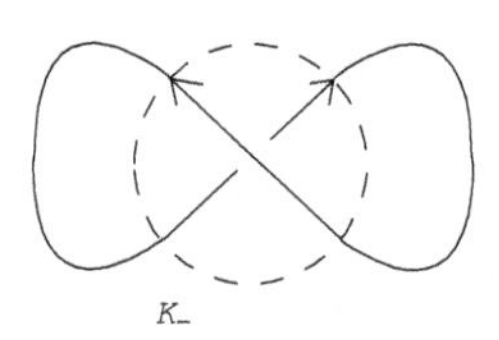

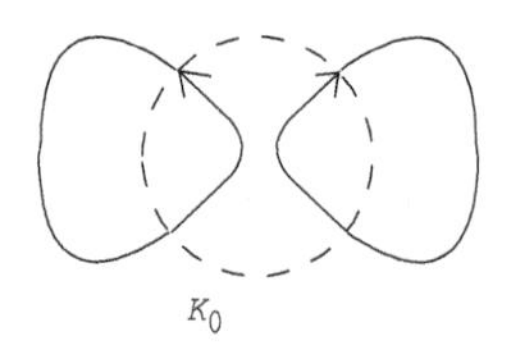

Figure 6.17 Three oriented links, K_+, K_-, and K_0. This figure is the same as Figure 6.10.

Both K_+ and K_- are the trivial knot — that is, the trivial one-component link. Thus, we have

$$V_{K_+}(t) = 1, \quad V_{K_-}(t) = 1.$$

K_0 is the trivial 2-component link oriented as drawn in Figure 6.16. We want to calculate the Jones polynomial $V_{K_0}(t)$ for K_0; the calculation is as follows:

$$\begin{aligned}(t^{\frac{1}{2}} - t^{-\frac{1}{2}})V_{K_0}(t) &= t^{-1}V_{K_+}(t) - tV_{K_-}(t)\\ &= t^{-1}\cdot 1 - t\cdot 1\\ &= t^{-1} - t.\end{aligned}$$

We thus conclude that $V_{K_0}(t) = -(t^{\frac{1}{2}} + t^{-\frac{1}{2}})$.

Recall that there are four ways to orient the trivial 2-component link, and that the four oriented links are all isotopic. See page 85.

6.4.2 *The Jones polynomial of the Hopf link*

We now calculate the Jones polynomial of the Hopf link, endowing it with the orientation shown in Figure 6.18.

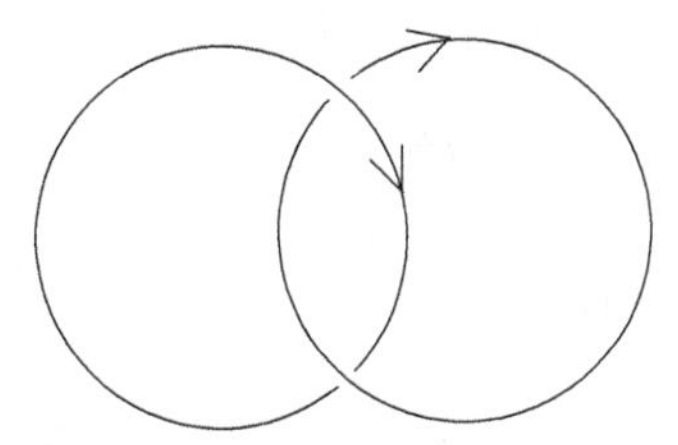

Figure 6.18 An oriented Hopf link. This figure is the same as Figure 6.11.

See Figure 6.19. We see three oriented links, K_+, K_-, and K_0, in the skein relation. By Definition 6.1, the Jones polynomials

$$V_{K_+}(t), V_{K_-}(t), V_{K_0}(t)$$

satisfy the relation

$$t^{-1}V_{K_+}(t) - tV_{K_-}(t) = (t^{\frac{1}{2}} - t^{-\frac{1}{2}})V_{K_0}(t).$$

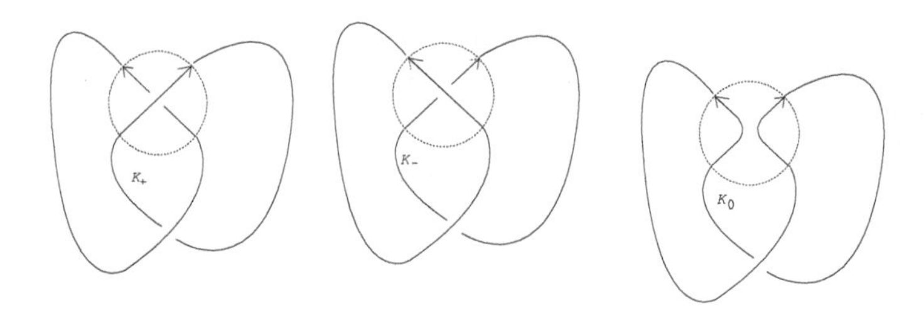

Figure 6.19 Three oriented links, K_+, K_-, and K_0. This figure is the same as Figure 6.12.

See Figure 6.19. K_- is isotopic to the trivial 2-component link. As we calculate in Section 6.4.1, the Jones polynomial $V_{K_-}(t)$ is $-(t^{\frac{1}{2}} + t^{-\frac{1}{2}})$. K_0 is the trivial knot. By Definition 6.1(1), the Jones polynomial $V_{K_0}(t)$ is 1.

Lastly, we have K_+, which in this case is the Hopf link oriented as drawn in Figure 6.18. We calculate the Jones polynomial $V_{K_+}(t)$ of K_+:

$$\begin{aligned} t^{-1}V_{K_+}(t) &= tV_{K_-}(t) + (t^{\frac{1}{2}} - t^{-\frac{1}{2}})V_{K_0}(t) \\ &= -t(t^{\frac{1}{2}} + t^{-\frac{1}{2}}) + (t^{\frac{1}{2}} - t^{-\frac{1}{2}}) \cdot 1. \end{aligned}$$

Thus, $V_{K_+}(t) = -t^{\frac{1}{2}} - t^{\frac{5}{2}}$.

6.4.3 *The Jones polynomial of the trefoil knot*

We now calculate the Jones polynomial of the trefoil knot oriented as drawn in Figure 6.20.

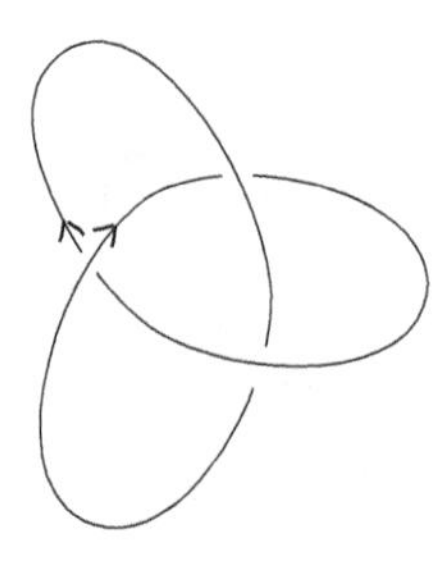

Figure 6.20 The trefoil knot. This figure is the same as Figure 6.13.

See Figure 6.21. We have three oriented links, K_+, K_-, and K_0, in the skein relation.

We employ Definition 6.1 to get that the Jones polynomials

$$V_{K_+}(t), V_{K_-}(t), V_{K_0}(t)$$

satisfy the relation

$$t^{-1}V_{K_+}(t) - tV_{K_-}(t) = (t^{\frac{1}{2}} - t^{-\frac{1}{2}})V_{K_0}(t).$$

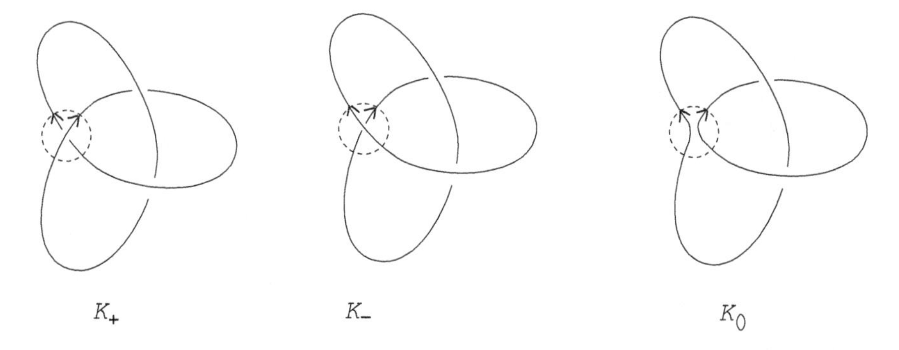

Figure 6.21 Three oriented links, K_+, K_-, and K_0. This figure is the same as Figure 6.14.

See Figure 6.21. K_- is the trivial knot. By Definition 6.1(1), the Jones polynomial $V_{K_-}(t)$ is 1. K_0 is the Hopf link oriented as drawn in Figure 6.18. As we calculate in Section 6.4.2, the Jones polynomial $V_{K_0}(t)$ of K_0 is $-t^{\frac{1}{2}} - t^{\frac{5}{2}}$.

K_+ is the trefoil knot in Figure 6.20. We now calculate the Jones polynomial $V_{K_+}(t)$ of K_+:

$$\begin{aligned} t^{-1}V_{K_+}(t) &= tV_{K_-}(t) + (t^{\frac{1}{2}} - t^{-\frac{1}{2}})V_{K_0}(t) \\ &= t \cdot 1 + (t^{\frac{1}{2}} - t^{-\frac{1}{2}})(-t^{\frac{1}{2}} - t^{\frac{5}{2}}) \\ &= t - (t + t^3 - 1 - t^2) \\ &= -t^3 + t^2 + 1. \end{aligned}$$

Therefore, we get the Jones polynomial $V_{K_+}(t) = t + t^3 - t^4$.

Now your turn! Calculate the Jones polynomial for the figure-eight knot shown again in Figure 6.22.

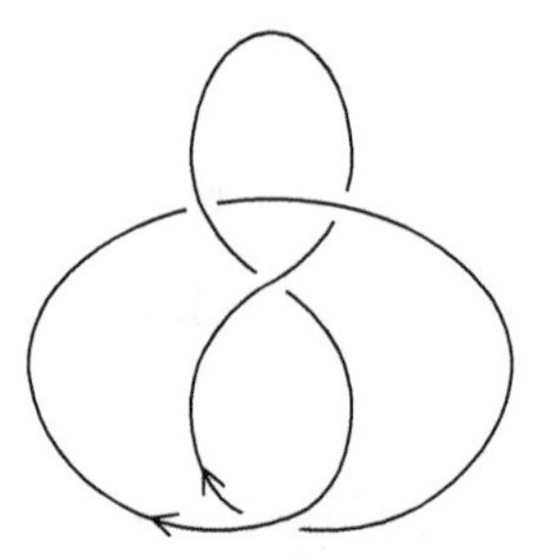

Figure 6.22 The figure-eight knot. This figure is the same as Figure 6.15.

Recall that the Alexander and Jones polynomials satisfy the relations

$$\Delta_{K_+}(t) - \Delta_{K_-}(t) = (t^{\frac{1}{2}} - t^{-\frac{1}{2}}) \cdot \Delta_{K_0}(t)$$

and

$$t^{-1}V_{K_+}(t) - tV_{K_-}(t) = (t^{\frac{1}{2}} - t^{-\frac{1}{2}})V_{K_0}(t),$$

respectively.

The reader may wonder: Is it possible to define your new polynomial for each isotopy type of links and a new identity as follows?

Given a link K of a given isotopy type, consider a polynomial $P_K(t)$, and fixed polynomials $a(t), b(t), c(t)$, satisfying:

$$a(t)P_{K_+}(t) - b(t)P_{K_-}(t) = c(t)P_{K_0}(t).$$

Here, we must have a unique P_K for a given link K. If two links J and K are isotopic, then $P_J = P_K$.

Moreover, considering arbitrary fixed $a(t)$, $b(t)$, and $c(t)$, they may not satisfy an identity among the three terms like the one given above.

You may change some condition. P_K may be more than a one-variable polynomial. P_K may be an equivalence class of polynomials. In fact, it is an open problem to list all possibilities for such identities among terms.

The HOMFLY-PT polynomial [24, 113] or Witten two-variable polynomial [132] is an example of knot polynomials, each of which satisfies a different identity from those for the Alexander and Jones polynomials.

In the high-dimensional case, the author [87] finds the same identity among three terms as those shown above, but also finds new identities [44], different from the Alexander and Jones identities in the 1-link case. We introduce them in Chapter 7.

6.5 The Jones and Alexander Polynomials and the Orientation of Knots and Links

6.5.1 *The Alexander polynomials of the positive and negative Hopf link*

We calculate the Alexander polynomial of the Hopf link, endowed with the orientation shown in Figure 6.23. In Figures 6.23 and 6.24, we show two different orientations of the Hopf link.

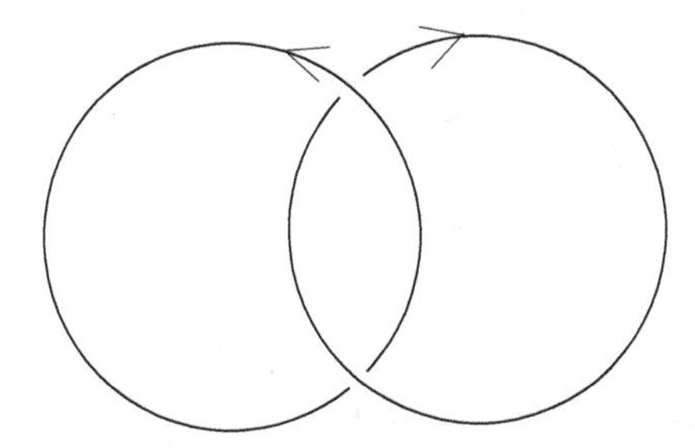

Figure 6.23 The Hopf link with an orientation. In this figure and Figure 6.24, we give different orientations.

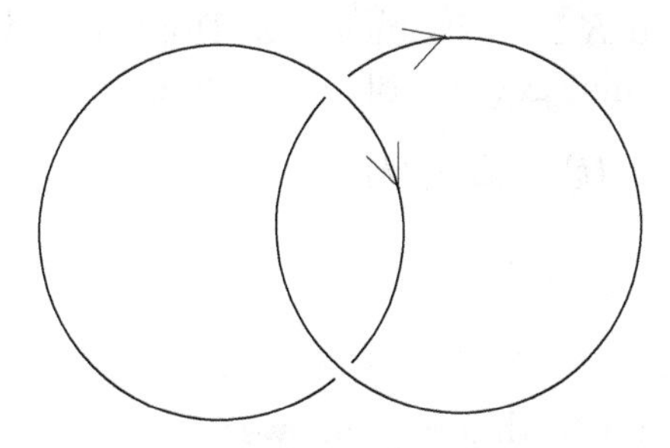

Figure 6.24 The Hopf link with an orientation different from that shown in Figure 6.23. This figure is the same as Figure 6.11.

We call the oriented link shown in Figure 6.23 the *negative Hopf link*, and that shown in Figure 6.24 the *positive Hopf link.* As we calculate in Section 6.3.2, the Alexander polynomial of the positive Hopf link is $t^{\frac{1}{2}} - t^{-\frac{1}{2}}$.

See the links pictured in Figure 6.25. The links, K_+, K_-, and K_0, are all oriented and in the skein relation. Recall Definition 6.1. The Alexander polynomials

$$\Delta_{K_+}(t), \Delta_{K_-}(t), \Delta_{K_0}(t)$$

satisfy the relation

$$\Delta_{K_+}(t) - \Delta_{K_-}(t) = (t^{\frac{1}{2}} - t^{-\frac{1}{2}}) \cdot \Delta_{K_0}(t).$$

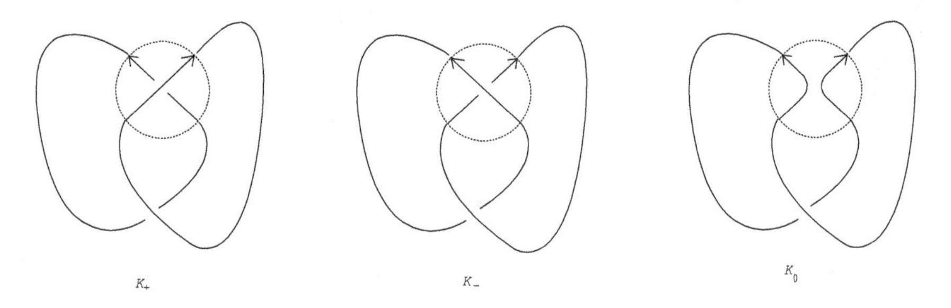

Figure 6.25 Three oriented links, K_+, K_-, and K_0.

In Figure 6.25, K_+ is the trivial 2-component link. The Alexander polynomial $\Delta_{K_+}(t)$ is 0 as we calculate in Section 6.3.1. K_0 is the trivial knot. By Definition 6.1(1), the Alexander polynomial $\Delta_{K_0}(t)$ is 1.

Lastly, we have K_- — the negative Hopf link. We calculate the Alexander polynomial $\Delta_{K_-}(t)$ of K_- as follows:

$$\begin{aligned}\Delta_{K_-}(t) &= \Delta_{K_+}(t) - (t^{\frac{1}{2}} - t^{-\frac{1}{2}})\Delta_{K_0}(t)\\ &= 0 - (t^{\frac{1}{2}} - t^{-\frac{1}{2}}) \cdot 1\\ &= -t^{\frac{1}{2}} + t^{-\frac{1}{2}}.\end{aligned}$$

Note that we got a different answer for the negative Hopf link from that for the positive Hopf link! We now calculate the Jones polynomials of the positive and negative orientations of the Hopf link.

6.5.2 *The Jones polynomials of the positive and the negative Hopf link*

The Jones polynomial of the positive Hopf link is $-t^{\frac{1}{2}} - t^{\frac{5}{2}}$ as we calculate in Section 6.4.2. So, let us calculate the Jones polynomial of the negative Hopf link shown in Figure 6.26.

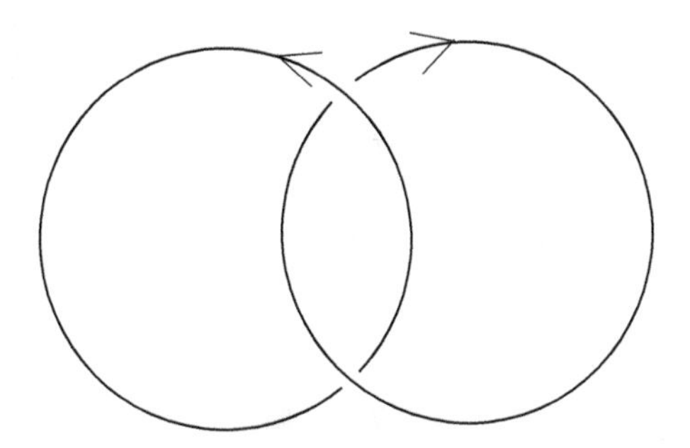

Figure 6.26 The negative Hopf link. This figure is the same as Figure 6.23.

See Figure 6.27. We have three links, K_+, K_-, K_0, in the skein relation. We again employ Definition 6.1, which tells us that the three Jones polynomials

$$V_{K_+}(t), V_{K_-}(t), V_{K_0}(t)$$

satisfy

$$t^{-1}V_{K_+}(t) - tV_{K_-}(t) = (t^{\frac{1}{2}} - t^{-\frac{1}{2}})V_{K_0}(t).$$

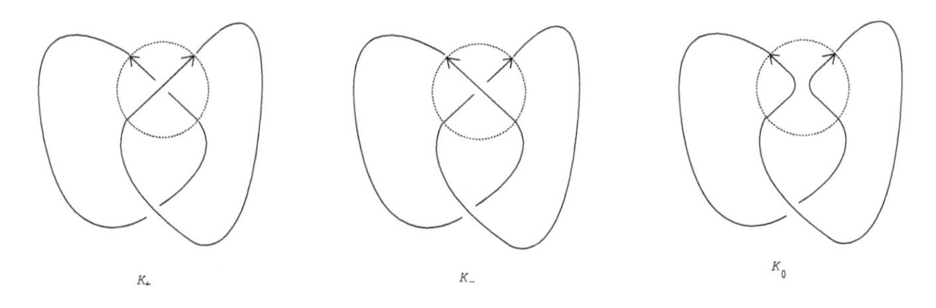

Figure 6.27 Three links, K_+, K_-, K_0, in the skein relation. This figure is the same as Figure 6.25.

In Figure 6.27, K_+ is the trivial 2-component link, whose Jones polynomial $V_{K_+}(t)$ is $-t^{\frac{1}{2}} - t^{\frac{5}{2}}$ as we calculate in Section 6.4.1. K_0 is the trivial knot. By Definition 6.1(1), the Jones polynomial $V_{K_0}(t)$ is 1.

K_- is the negative Hopf link. We want to calculate the Jones polynomial $V_{K_-}(t)$, which we can do using $V_{K_+}(t)$ and $V_{K_0}(t)$:

$$\begin{aligned} tV_{K_-}(t) &= t^{-1}V_{K_+}(t) - (t^{\frac{1}{2}} - t^{-\frac{1}{2}})V_{K_0}(t) \\ &= t^{-1}\{-(t^{\frac{1}{2}} + t^{-\frac{1}{2}})\} - (t^{\frac{1}{2}} - t^{-\frac{1}{2}}) \\ &= -t^{-\frac{3}{2}} - t^{\frac{1}{2}}. \end{aligned}$$

Therefore,

$$V_{K_-}(t) = -t^{-1}(t^{-\frac{3}{2}} - t^{\frac{1}{2}}) = -t^{-\frac{5}{2}} - t^{-\frac{1}{2}}.$$

We thus see that the Jones polynomial of the negative Hopf link is different from that of the positive Hopf link, just as it was for the Alexander polynomial.

6.6 Mirror Images of Knots and Links

Consider a diagram of any knot or link — call it L — and change all the crossings as we do to the knot shown in Figure 6.28. The resulting diagram represents a new knot or link, called the *mirror image* of L. We often use L^* to denote the mirror image of L.

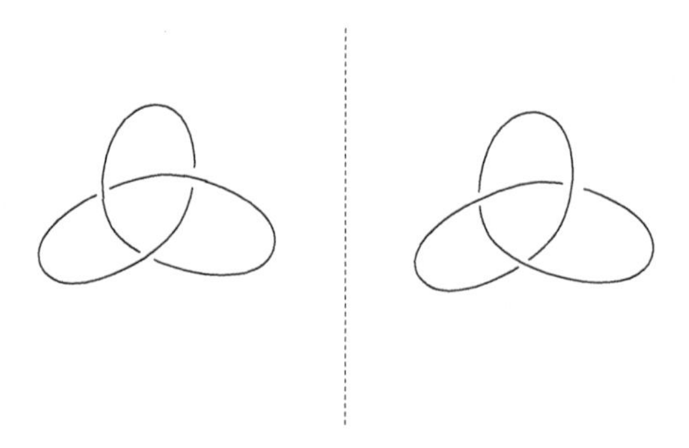

Figure 6.28 A mirror image of a knot.

Figure 6.29 helps elucidate why we call this new knot or link with opposite crossings the "mirror image." Note that our original diagram of L lies in some plane P, which we can regard as a mirror. So when we say "mirror image," we mean the mirror image with respect to the plane P.

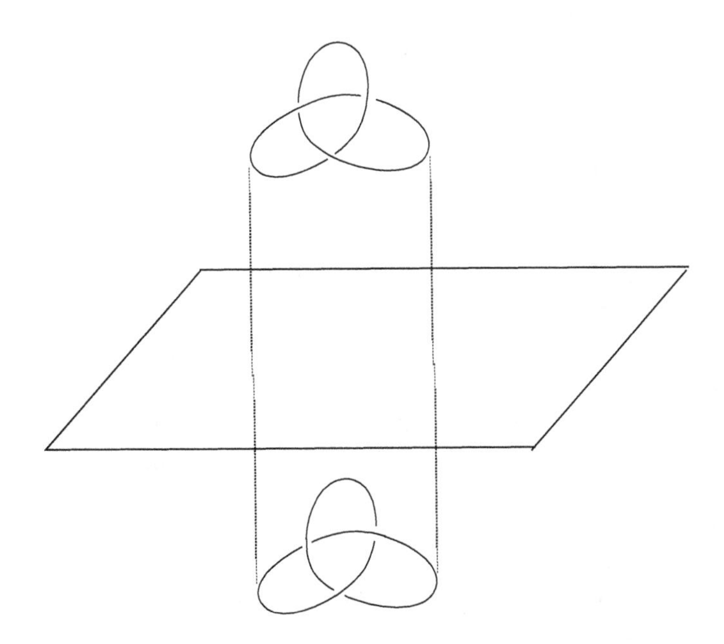

Figure 6.29 Why we say "mirror images."

In Figure 6.28, one knot is the mirror image of the other. The knot shown on the right is called the *right-hand trefoil knot*, and that on the left, the *left-hand trefoil knot*.

Now, let L be an oriented link, where the orientation is represented by arrows in the diagram. In the diagram of L, flip all the crossings as shown in Figure 6.30. We get a new link diagram, but note that the orientation arrows did not change. Thus, our new diagram actually determines a new oriented link. We will call this new oriented link the *mirror image* of L, and again use L^* to denote it.

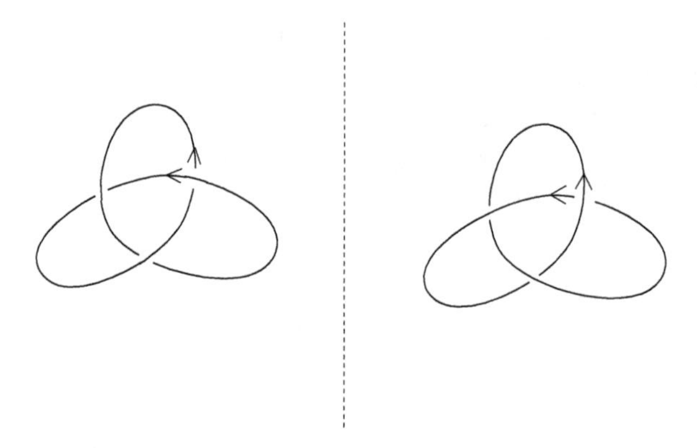

Figure 6.30 A mirror image of an oriented knot.

Let K be the right-handed (or, left-handed) trefoil knot with an orientation. Then K and $-K$ are isotopic. Recall an isotopy in Figure 6.4.

Note that in Figure 6.13, as well as in Figure 6.20, we draw the right-hand trefoil knot. Thus, it is specifically the right-hand trefoil knot whose Alexander and Jones polynomials we are calculating in Sections 6.3.3 and 6.4.3, respectively.

It is natural to ask the following question:

Question 6.4. Is the right-handed trefoil knot isotopic to the left-handed trefoil knot?

We answer this question in the following section.

Let L be an oriented link and consider the mirror image L^* of L. Change the orientation of the mirror image L^* by flipping the orientation of each component and denote the resulting link $-L^*$. See Figure 6.31 for an example.

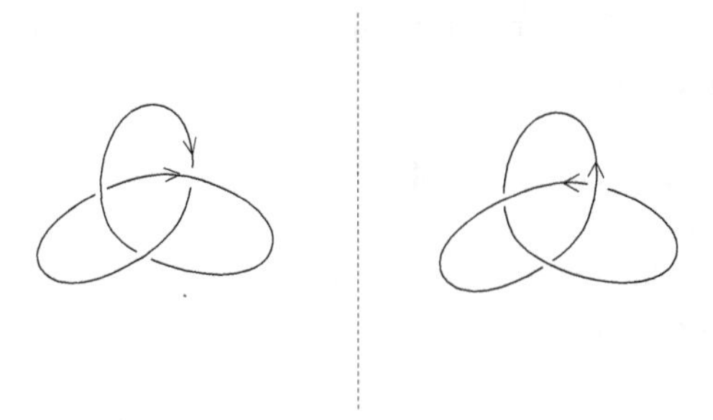

Figure 6.31 When the left knot is denoted by K, the right knot is usually denoted by $-K^*$.

6.6.1 *The Jones polynomial of the right-hand (or left-hand) trefoil knot*

From Section 6.4.3, we have that the Jones polynomial of the right-hand trefoil knot is $t+t^3-t^4$. In this section, we calculate the Jones polynomial of the left-hand trefoil knot. In doing so, we will answer Question 6.4.

See Figure 6.32. The three Jones polynomials

$$V_{K_+}(t), V_{K_-}(t), V_{K_0}(t)$$

satisfy the relation

$$t^{-1}V_{K_+}(t) - tV_{K_-}(t) = (t^{\frac{1}{2}} - t^{-\frac{1}{2}})V_{K_0}(t).$$

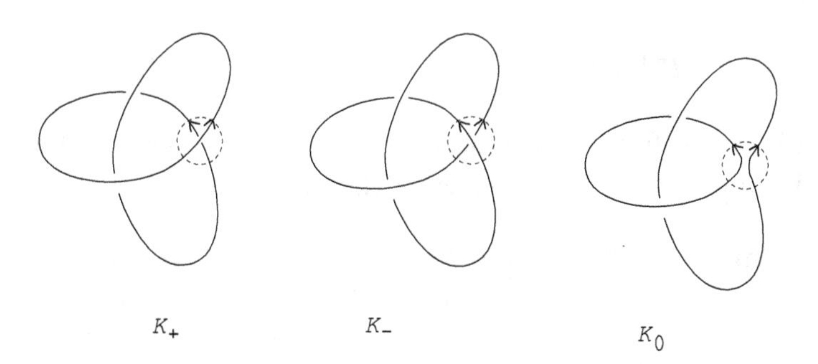

Figure 6.32 An example of the skein relation.

In Figure 6.32, K_+ is the trivial knot and K_0 is the negative Hopf link, whose Jones polynomials we know to be

$$V_{K_+}(t) = 1, \quad V_{K_0}(t) = -t^{-\frac{5}{2}} - t^{-\frac{1}{2}}.$$

We thus calculate the Jones polynomial $V_{K_-}(t)$ of K_- — the left-hand trefoil — as follows:

$$\begin{aligned} tV_{K_-}(t) &= t^{-1}V_{K_+}(t) - (t^{\frac{1}{2}} - t^{-\frac{1}{2}})V_{K_0}(t) \\ &= t^{-1} \cdot 1 - (t^{\frac{1}{2}} - t^{-\frac{1}{2}})(-t^{-\frac{5}{2}} - t^{-\frac{1}{2}}) \\ &= t^{-1} + (t^{\frac{1}{2}} - t^{-\frac{1}{2}})(t^{-\frac{5}{2}} + t^{-\frac{1}{2}}) \\ &= t^{-1} + (t^{-2} + 1 - t^{-3} + (t^{-1}) \\ &= t^{-2} + 1 - t^{-3}. \end{aligned}$$

We get the polynomial

$$V_{K_-}(t) = t^{-1} + t^{-3} - t^{-4}.$$

The Jones polynomial, then, of the right-hand and left-hand trefoil knots are $t + t^3 - t^4$ and $t^{-1} + t^{-3} - t^{-4}$, respectively. We established that the Jones polynomial is a knot invariant, so we conclude that the right-hand trefoil and the left-hand trefoil are not in fact isotopic.

However, note the following. Try substituting t^{-1} for t in the Jones polynomial of the the right-hand trefoil knot. We get:

$$t^{-1} + (t^{-1})^3 - (t^{-1})^4 = t^{-1} + t^{-3} - t^{-4},$$

which is precisely the Jones polynomial of the left-hand trefoil knot. This is an example of the following theorem in action.

Theorem 6.5 ([32]). *Let L be an oriented link and let L^* denote its mirror image. Let $V_L(t)$ and $V_{L^*}(t)$ be the Jones polynomials of L and L^*, respectively. We have the following identity:*

$$V_L(t^{-1}) = V_{L^*}(t).$$

6.6.2 *The Alexander polynomial of the right-hand (respectively, the left-handed) trefoil knot*

In Section 6.3.3, we calculate the Alexander polynomial of the right-hand trefoil knot to be $t - 1 + t^{-1}$. In this section, we calculate the Alexander polynomial of the left-hand trefoil knot.

See Figure 6.33. The three Alexander polynomials

$$\Delta_{K_+}(t), \Delta_{K_-}(t), \Delta_{K_0}(t)$$

of the knots shown satisfy

$$\Delta_{K_+}(t) - \Delta_{K_-}(t) = (t^{\frac{1}{2}} - t^{-\frac{1}{2}}) \cdot \Delta_{K_0}(t).$$

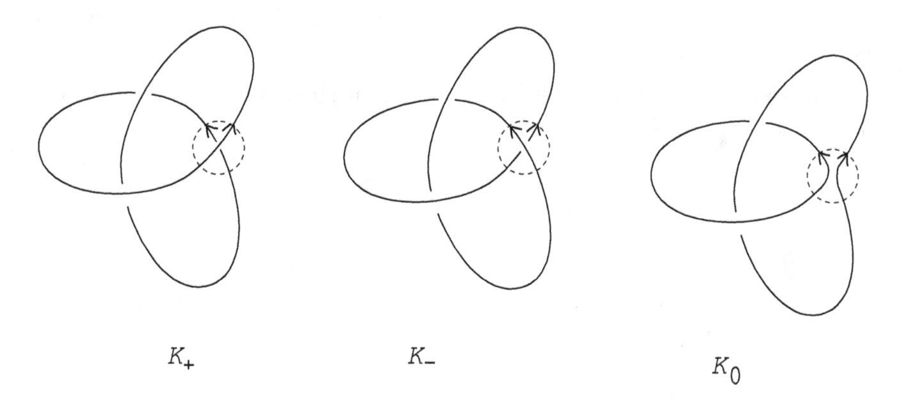

Figure 6.33 An example of the skein relation. This figure is the same as Figure 6.32.

K_+ is the trivial knot and K_0 is the negative Hopf link. Hence,

$$\Delta_{K_+}(t) = 1, \quad \Delta_{K_0}(t) = -t^{\frac{1}{2}} + t^{-\frac{1}{2}}.$$

Therefore, the Alexander polynomial $\Delta_{K_-}(t)$ of the knot K_- — the left-hand trefoil — is calculated as follows.

$$\begin{aligned}\Delta_{K_-}(t) &= \Delta_{K_+}(t) - (t^{\frac{1}{2}} - t^{-\frac{1}{2}}) \cdot \Delta_{K_0}(t)\\ &= 1 - (t^{\frac{1}{2}} - t^{-\frac{1}{2}})(-t^{\frac{1}{2}} + t^{-\frac{1}{2}})\\ &= t - 1 + t^{-1}.\end{aligned}$$

Thus, we see that the left- and right-hand trefoil knots have the same Alexander polynomial $t - 1 + t^{-1}$.

We saw in Section 6.6.1 that the left- and right-hand trefoil knots are not isotopic, yet the two knots have the same Alexander polynomial.

So, in the case of the trefoil, the Jones polynomial can distinguish between the knot and its mirror image, but the Alexander polynomial can't. We have two non-isotopic knots whose Alexander polynomials are the same, but whose Jones polynomials are different.

Note that there also exist non-isotopic knots with the same Jones polynomial but different Alexander polynomials. Hence, sometimes the Alexander polynomial is useful for distinguishing between two knots, and other times not so much, and we need the Jones polynomial. The same is true vice versa. Consequently, it is important to keep both of these polynomials in mind.

We explain high-dimensional versions of these kinds of local move identities in Chapter 7.

Chapter 7

Local Moves and Knot Polynomials in Higher Dimensions

7.1 The Alexander Polynomials of High-Dimensional Knots

We introduce a general definition of the Alexander polynomial for n-dimensional knots for any positive integer $n \geqq 1$. This definition in the $n = 1$ case is the same as the definition in the 1-link case introduced in the previous sections. (See [39, 117].)

For now, it is enough for the beginners to understand that we assign to each n-knot in $\mathbb{R}^{n+2}$ a polynomial. So they may skip ahead.

See [1, 54, 60] for details on this and [74] to read about homology groups.

We need to introduce $\mathbb{Q}[t, t^{-1}]$-balanced classes of polynomials.

Definition 7.1. Two polynomials $f(t), g(t) \in \mathbb{Q}[t, t^{-1}]$ are said to be $\mathbb{Q}[t, t^{-1}]$-*balanced* if there is an integer n and a non-zero rational number r such that $f(t) = r \cdot t^n \cdot g(t)$.

Definition 7.2. Let n be any positive integer, $n \geqq 1$, and let K be an (oriented) n-knot in S^{n+2}. Let $N(K)$ be a tubular neighborhood of K in S^{n+2}, where $N(K) = K\times$(the open 2-disc). Let $X = S^{n+2} - K$, the complement of K in S^{n+2}. Take the cyclic covering space $\widetilde{X}$ of X, defined naturally by using the orientation of K and that of S^{n+2}. Then $H_i(\widetilde{X}; \mathbb{Q})$ is regarded as a $\mathbb{Q}[t, t^{-1}]$-module.

According to module theory, it holds that any $\mathbb{Q}[t, t^{-1}]$-module is congruent to

$$(\mathbb{Q}[t, t^{-1}]/\lambda_1) \oplus \cdots \oplus (\mathbb{Q}[t, t^{-1}]/\lambda_l) \oplus (\oplus^k \mathbb{Q}[t, t^{-1}]),$$

where we have the following:

(1) $\lambda_* \in \mathbb{Q}[t, t^{-1}]$ is not zero,
(2) λ_* is not the $\mathbb{Q}[t, t^{-1}]$-balanced class of 1, and
(3) k is the rank of the free part.

Let $H_i(\widetilde{X}; \mathbb{Q})$ be as above. Then the $\mathbb{Q}[t, t^{-1}]$-*i*-*Alexander polynomial* is the $\mathbb{Q}[t, t^{-1}]$-balanced class of

$$\begin{cases} \lambda_1 \cdot \ldots \cdot \lambda_l & \text{if } k = 0 \text{ and } H_i(\widetilde{X}; \mathbb{Q}) \text{ is non-trivial} \\ 0 & \text{if } k \neq 0 \\ 1 & \text{if } H_p(\widetilde{X}; \mathbb{Q}) \cong 0. \end{cases}$$

Remark: If K may be a (not necessarily connected) n-dimensional closed oriented manifold, which may not be the n-sphere, then the $\mathbb{Q}[t, t^{-1}]$-*i*-*Alexander polynomial* is defined in the same fashion. See [74] for background on manifolds.

We define an *n-dimensional manifold*, or *n-manifold*, M, as follows: For a space M, take any point $p \in M$. There exists a *neighborhood* $N(p)$ in M — that is, a subset of M near p — such that $N(p)$ is *homeomorphic* to $\mathbb{R}^n$ (informally: $N(p)$ "acts like" $\mathbb{R}^n$). The terms "neighborhood" and "homeomorphic" are math ones. The sphere S^2 and the torus $S^1 \times S^1$ are examples of two-dimensional manifolds. The n-sphere S^n is an n-dimensional manifold.

Remark: While we have introduced here a high-dimensional version of the Alexander polynomial, it is still an open problem to find an analog of the Jones polynomial for high-dimensional knots. See Section 7.5.

7.2 Local Move Identities for 2-Knots

Let K_+ be a 2-knot in $\mathbb{R}^4$. Assume that K_+, K_-, and K_0 differ only in the 4-ball B^4 in $\mathbb{R}^4$. They intersect B^4 as shown in Figures 7.1–7.3. Note that Figure 7.1 (respectively, Figure 7.2) is the same as Figure 3.1 (respectively, 3.2) of Chapter 3. In this scenario, we say that (K_+, K_-, K_0) makes a *ribbon-move triple*.

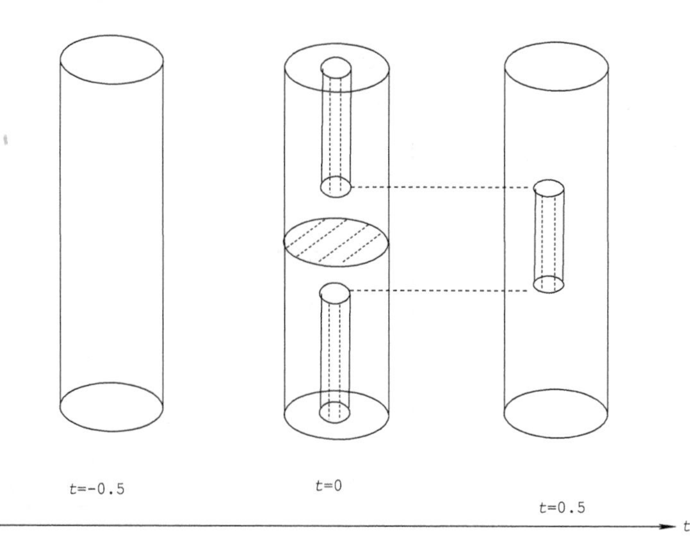

Figure 7.1 Figures 7.1, 7.2, and 7.3 make the ribbon-move triple.

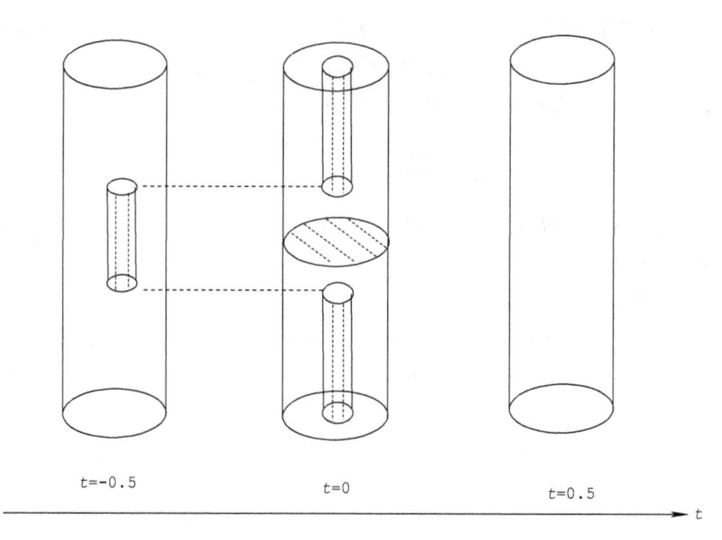

Figure 7.2 Figures 7.1, 7.2, and 7.3 make the ribbon-move triple.

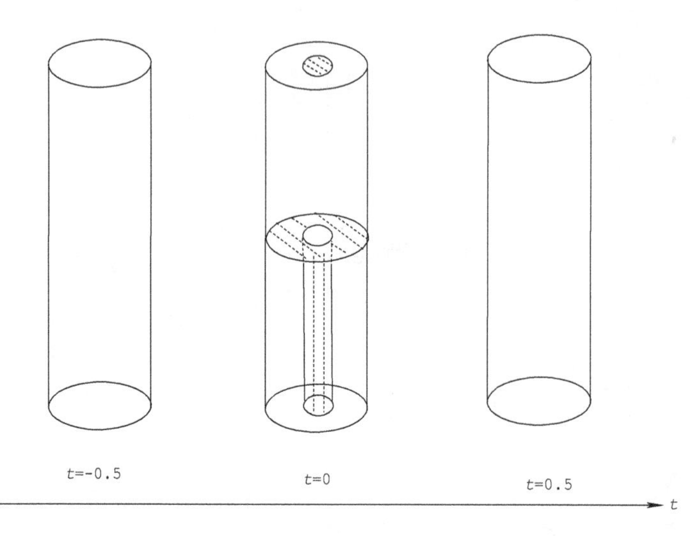

Figure 7.3 Figures 7.1, 7.2, and 7.3 make the ribbon-move triple.

The author proved the following theorem.

Theorem 7.1 ([87]). *Let K_+ and K_- be 2-knots in $\mathbb{R}^4$ and let K_0 be a two-dimensional submanifold (see [74]) of $\mathbb{R}^4$. Suppose that (K_+, K_-, K_0) is a ribbon-move triple. Then we have*

$$\Delta^p_{K_+} - \Delta^p_{K_+} = (t-1) \cdot \Delta^p_{K_0},$$

where Δ^p_K is a polynomial whose balanced class is the p-Alexander polynomial for K.

Remark: K_0 is S^2 in some cases and is not in the other cases. Theorem 7.1 is true both if K_0 is S^2 and if K_0 is not S^2.

7.3 Local Move Identities and (p, q)-Pass-Move on $(p+q-1)$-Knots

We introduce a similar result to that given in Section 7.2 but for higher-dimensional knots. Let K_+ be an n-knot in $\mathbb{R}^{n+2}$. Assume that K_+, K_-, and K_0 differ only in the $(n+2)$-ball B^{n+2} in $\mathbb{R}^{n+2}$. They intersect B^{n+2} as shown in Figures 7.4–7.6. Note that Figure 7.4 (respectively, 7.5) is the same as Figure 4.26 (respectively, 4.27) of

Chapter 4. Then we say that (K_+, K_-, K_0) makes a *(p,q)-(pass)-move triple.*

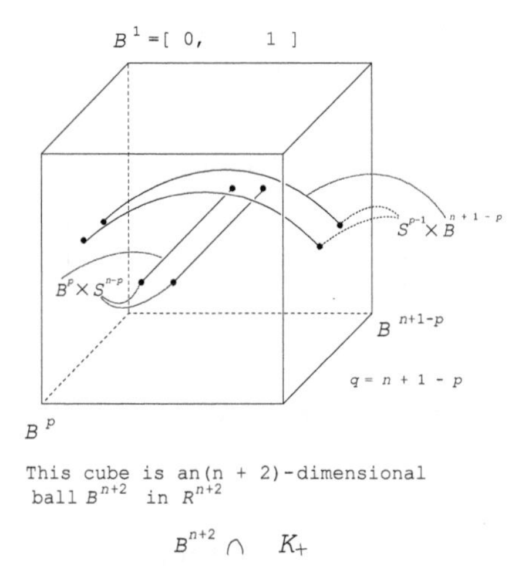

Figure 7.4 Figures 7.4, 7.5, and 7.6 make the (p,q)-(pass-)move-triple.

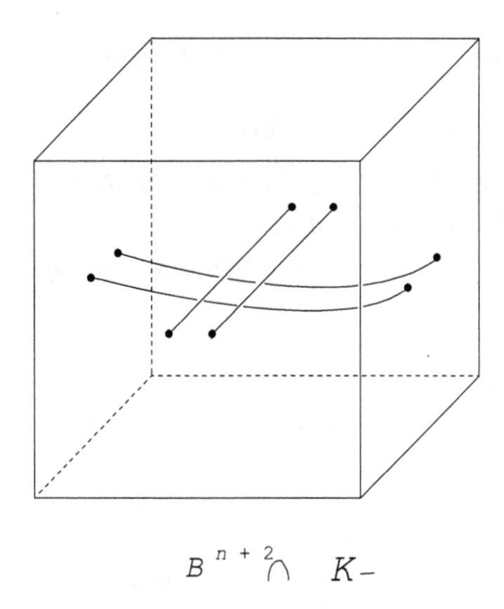

Figure 7.5 Figures 7.4, 7.5, and 7.6 make the (p,q)-(pass-)move-triple.

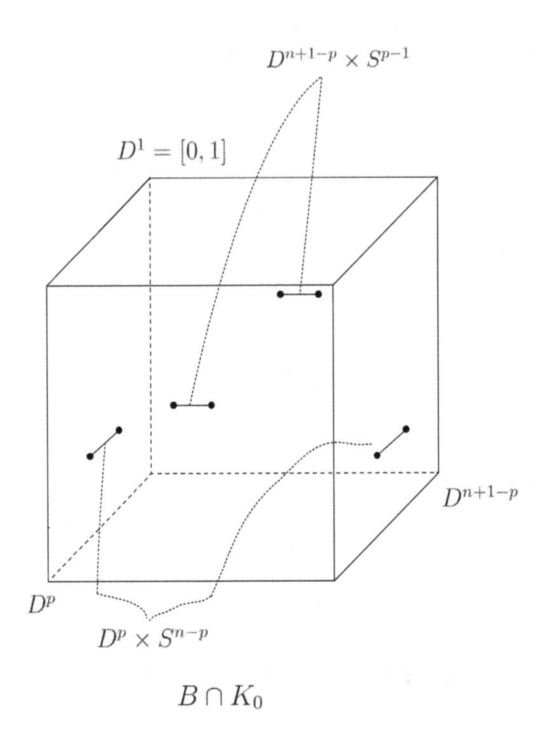

Figure 7.6 Figures 7.4, 7.5, and 7.6 make the (p, q)-(pass-)move-triple.

The author proved the following.

Theorem 7.2 ([87]). *Let K_+, K_- be n-knots in S^{n+2} and let K_0 be an n-dimensional submanifold of S^{n+2}. Suppose that (K_+, K_-, K_0) is a (p, q)-pass-move-triple, where $p \neq q$ and $p + q + 1 = n$. Then we have*

$$\Delta^p_{K_+} - \Delta^p_{K_+} = (t - 1) \cdot \Delta^p_{K_0},$$

where Δ^p_K is a polynomial whose balanced class is the p-Alexander polynomial for K.

Remark: K_0 is S^{n+2} in some cases and is not in the other cases. Theorem 7.2 is true both if K_0 is S^{n+2} and if K_0 is not S^{n+2}.

When $p = 1$, $q = 2$, and $n = 4$, the (p, q)-move-triple is drawn as in Figures 7.7–7.9. Note that Figure 7.7 (respectively, Figure 4.29) is the same as Figure 4.28 (respectively, Figure 4.29) of Chapter 4.

The author [85] found a relation between the ribbon-move triple and the (1, 2) move triple, which satisfy the same identity.

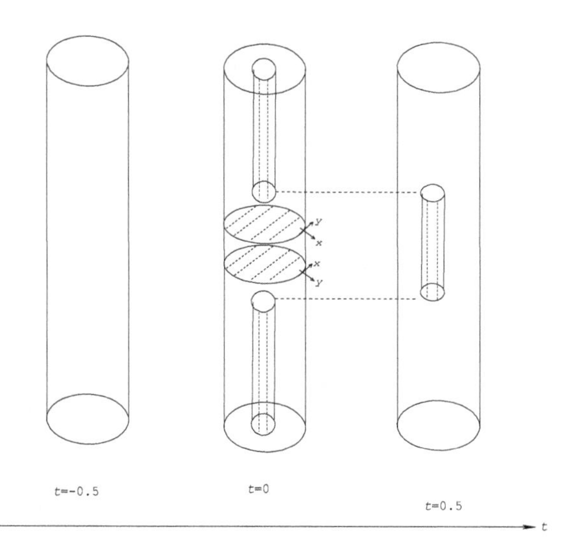

Figure 7.7 Figures 7.7, 7.8 and 7.9 make the (1, 2)-pass-move triple.

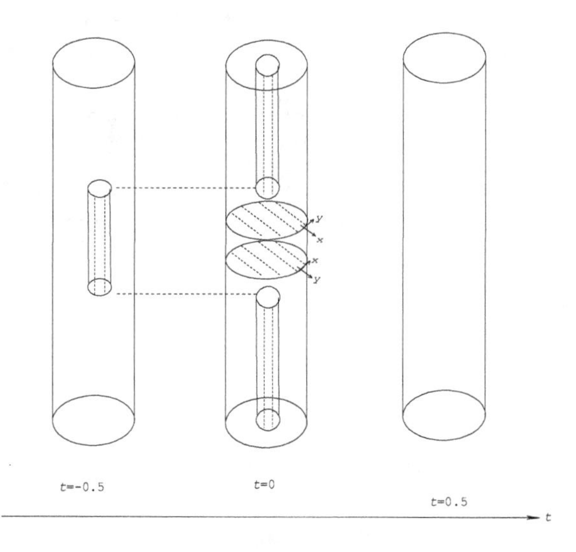

Figure 7.8 Figures 7.7, 7.8 and 7.9 make the (1, 2)-pass-move triple.

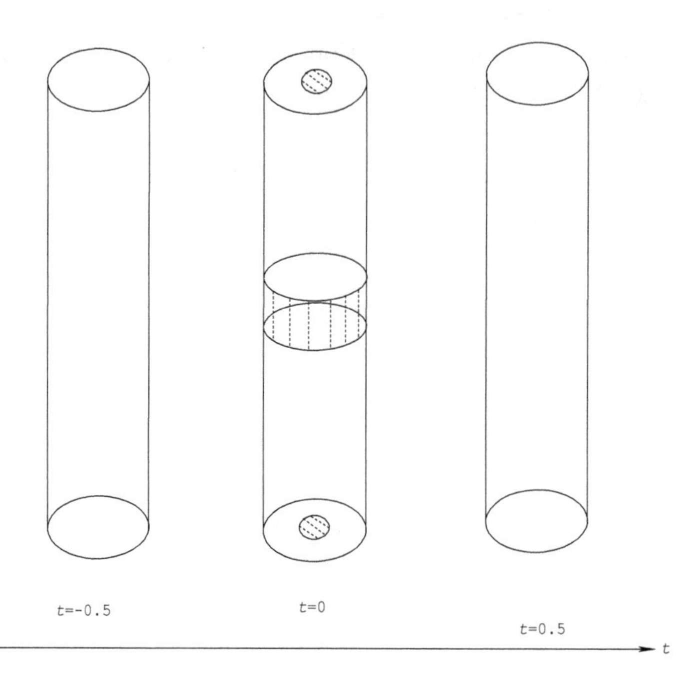

Figure 7.9 Figures 7.7, 7.8 and 7.9 make the (1, 2)-pass-move triple.

7.4 Local Move Identities and the Twist-Move on $(2m+1)$-Knots

We now introduce another high-dimensional analogue of a familiar local move. Let K_+ be a $(2m+1)$-knot in $\mathbb{R}^{2m+3}$. Assume that K_+, K_-, and K_0 differ only in the $(2m+2)$-ball B^{2m+2} in $\mathbb{R}^{2m+4}$. They intersect B^{2m+2} as shown in Figures 7.10–7.12. Note that Figure 7.10 (respectively, Figure 7.11) is the same as Figure 4.37 (respectively, Figure 4.38) of Chapter 4. In this case, we say that (K_+, K_-, K_0) makes a *twist-move triple.*

The author proved the following result.

Theorem 7.3 ([44, 87]). *Let K_+ be a $(2p+1)$-knot in $\mathbb{R}^{2p+3}$, where p is a non-negative integer, and let K_- and K_0 be n-dimensional submanifolds of S^{n+2}. Suppose that (K_+, K_-, K_0) is related by the twist-move, where $p \neq q$. Then we have*

$$\Delta^p_{K_+} - \Delta^p_{K_+} = (t - (-1)^p) \cdot \Delta^p_{K_0},$$

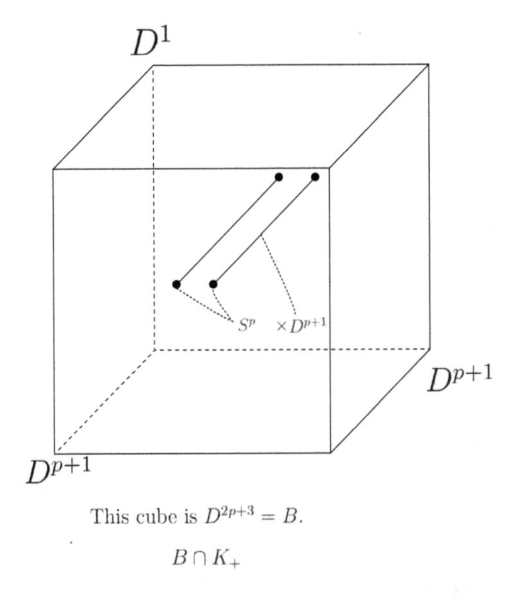

Figure 7.10 Figures 7.10, 7.11, and 7.12 make the twist-move triple.

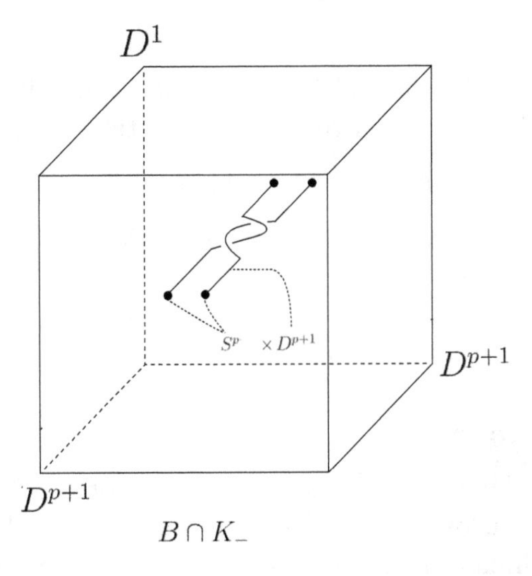

Figure 7.11 Figures 7.10, 7.11, and 7.12 make the twist-move triple.

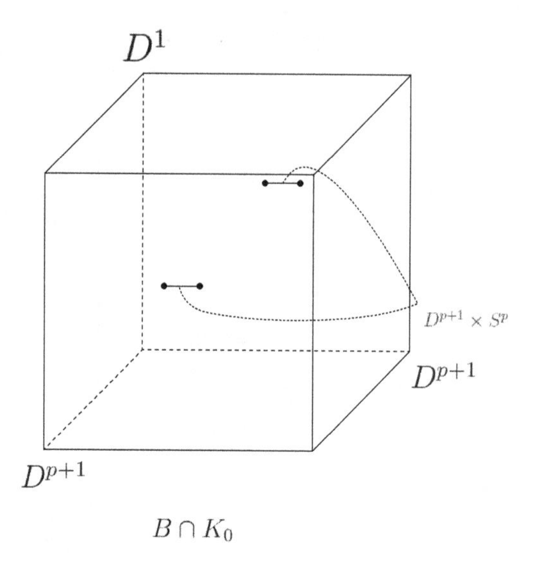

Figure 7.12 Figures 7.10, 7.11, and 7.12 make the twist-move triple.

where Δ_K^p *is a polynomial whose balanced class is the p-Alexander polynomial for K.*

Note $(t-(-1)^p)$ in the right-hand side of the identity. It is different from the other identities in the previous sections.

Remark: For 3-knots, we have two kinds of identities:

$$\Delta_{K_+}^1 - \Delta_{K_+}^1 = (t-1) \cdot \Delta_{K_0}^1,$$

and

$$\Delta_{K_+}^2 - \Delta_{K_+}^2 = (t+1) \cdot \Delta_{K_0}^2.$$

7.5 Knot Products

Let K be an n-knot in S^{n+2} and J be an m-knot in S^{m+2}. Their *knot product* $K \otimes J$ yields an $(n+m+3)$-knot in S^{n+m+5}. This concept is given by Kauffman in [42], and after that, Kauffman and Neumann in [36].

Kauffman and the author [44–46] have proved several results on knot products. Two of them are the following identities.

Theorem 7.4 ([44–46]). *Let K_+, K_-, and K_0 satisfy the one- or high-dimensional skein relation (introduced in the previous sections).*

(1) *Let $\mu \in \mathbb{N} \cup \{0\}$. There is a polynomial $\Delta_{2\mu+1}(K_* \otimes^\mu (\text{Hopf})) \in \mathbb{Q}[\mathrm{t}, \mathrm{t}^{-1}]$ whose $\mathbb{Q}[t, t^{-1}]$-balanced class is the $(2\mu+1)$-$\mathbb{Q}[t, t^{-1}]$-Alexander polynomial $A_{2\mu+1}(K_* \otimes^\mu (\text{Hopf})$ $(* = +, -, 0)$ such that*

$$\Delta_{2\mu+1}(K_+ \otimes^\mu (\text{Hopf}) - \Delta_{2\mu+1}(\mathrm{K}_- \otimes^\mu (\text{Hopf})$$
$$= (t-1) \cdot \Delta_{2\mu+1}(K_0 \otimes^\mu (\text{Hopf})),$$

where (Hopf) *denotes the Hopf link.*

(2) *Let $\nu \in \mathbb{N} \cup \{0\}$. There is a polynomial $\Delta_{\nu+1}(K_* \otimes^\nu [2]) \in \mathbb{Q}[t, t^{-1}]$ whose $\mathbb{Q}[t, t^{-1}]$-balanced class is the $(\nu+1)$-$\mathbb{Q}[t, t^{-1}]$-Alexander polynomial $A_{\nu+1}(K_* \otimes^\nu [2])$ $(* = +, -, 0)$ such that*

$$\Delta_{\nu+1}(K_+ \otimes^\nu [2]) - \Delta_{\nu+1}(K_- \otimes^\nu [2])$$
$$= (t + (-1)^{\nu+1}) \cdot \Delta_{\nu+1}(K_0 \otimes^\nu [2]),$$

where [2] *denotes the empty knot of degree two (introduced in* [36, 42]*).*

Chapter 8

Important Topics in Knot Theory

8.1 Slice Knots in $\mathbb{R}^3$

We introduce *slice knots.*

Definition 8.1. Let K be a knot in $\mathbb{R}^3$. Let $\mathbb{R}^4_{\geqq 0} = \{(x, y, z, t)|x, y,$ and z are arbitrary numbers, $t \geqq 0\}$. Regard $\mathbb{R}^3$ as $\{(x, y, z, 0)|x, y,$ and z are arbitrary numbers$\}$ in $\mathbb{R}^4_{\geqq 0}$, that is, $\mathbb{R}^3$ at $t = 0$ in $\mathbb{R}^4_{\geqq 0}$. We assume that K is in $\mathbb{R}^3$ at $t = 0$ in $\mathbb{R}^4_{\geqq 0}$. We say that K is a *slice knot* if there exists a disc D embedded in $\mathbb{R}^4_{\geqq 0}$ whose boundary ∂D is K in $\mathbb{R}^3$ at $t = 0$.

By virtue of the methods we outline in Section 2.2 of Chapter 2, we have the following. Note that we use $\mathbb{R}^4_{\geqq 0}$ in Definition 8.1 and that we use $\mathbb{R}^4$ in Fact 8.2.

Fact 8.2 (Known). Let K be an arbitrary knot in $\mathbb{R}^3$. Recall $\mathbb{R}^4 = \{(x, y, z, t)|x, y, z,$ and t are arbitrary numbers$\}$. Regard $\mathbb{R}^3$ as $\{(x, y, z, 0)|x, y,$ and z are arbitrary numbers$\}$ in $\mathbb{R}^4$, that is, $\mathbb{R}^3$ at $t = 0$ in $\mathbb{R}^4$. We assume that K is in $\mathbb{R}^3$ at $t = 0$ in $\mathbb{R}^4$. Then there is an embedded disc D in $\mathbb{R}^4$ whose boundary ∂D is K in $\mathbb{R}^3$ at $t = 0$.

On the other hand, we have the following theorem.

Theorem 8.3 (Known). *There is a non-slice knot K in $\mathbb{R}^3$.*

The trefoil knot is an example of non-slice knots. The trivial knot, however, is most definitely slice.

It is natural to wonder: Is there a non-trivial knot which is slice? We have the following theorem which answers this question.

Theorem 8.4 (Known). *There is a non-trivial knot K in $\mathbb{R}^3$ which is slice.*

Proof of Theorem 4. Proof by picture. See Figure 8.1. □

Recall the notation $-L^*$ in Question 6.4 in Chapter 6 and the connected sum $J\sharp K$ introduced in Section 5.3 of Chapter 5. For any non-trivial 1-knot A, the connected sum $A\sharp(-A^*)$ is a slice and non-trivial knot.

We have an open question pertaining to slice knots.

Question 8.5. Given an arbitrary knot K, is K a slice knot? More rather, give a characterization of all knots K which are slice.

Question 8.5 actually has two cases depending on the category we are working in — the smooth category or the topological category — when we interpret the question. Moreover, while we don't have a complete answer to Question 8.5, we at least know that the smooth category and the topological category give different answers.

Donaldson's exciting work [13] has an immediate corollary (first observed by Akbulut and Casson, and appears in Cochran and Gompf's paper [8]): there are classical knots with trivial Alexander polynomials which are not smoothly slice. A year after Donaldson's work, Freedman [22], and Freedman with Quinn [23], proved a four-dimensional topological surgery theorem for manifolds with fundamental group $\mathbb{Z}$. This implies that a knot with trivial Alexander polynomial is in fact topologically (locally flat) slice.

It is known that all 1-knots bound an immersed 2-disc in $\mathbb{R}^3$. Let K be a knot. We call K a *ribbon* knot if K bounds an immersed 2-disc such that all singularities are of the form illustrated in Figure 8.2.

Two kinds of bold dotted lines are drawn.

(1) The bolder one of the two is a ribbon singularity.
(2) The thinner ones are parts of a knot.

The bold lines, which are as bold as the bold dotted lines are parts of a knot.

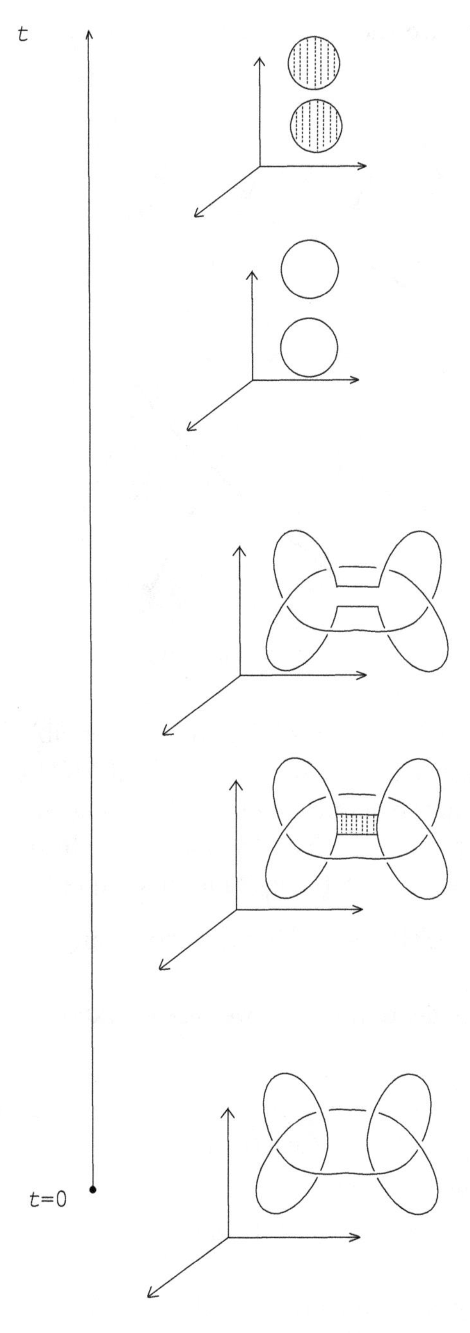

Figure 8.1 A slice knot.

Shaded parts are those of an immersed disc.

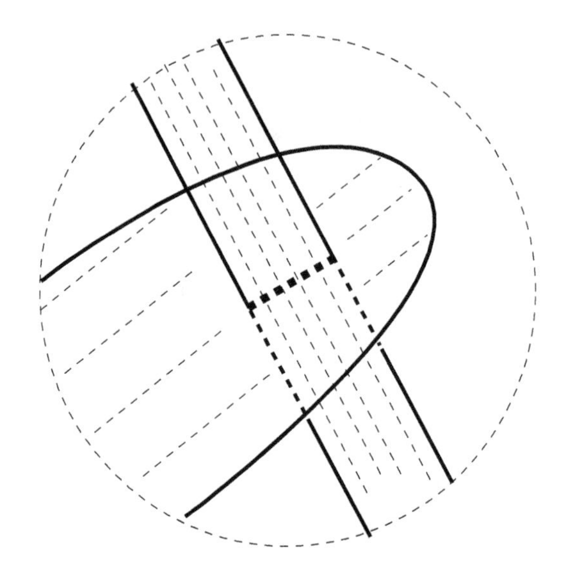

Figure 8.2 Ribbon singularity.

The knot in $\mathbb{R}^3$ at $t = 0$ of Figure 8.1 is a ribbon knot which is non-trivial. For any 1-knot K, the knot $K\sharp(-K^*)$ is a ribbon knot.

It follows from the definition of a ribbon knot that all ribbon knots are slices in both the smooth and topological categories. However, the converse in the smooth category is an open question:

Question 8.6. In the smooth category, is any slice knot a ribbon knot?

In the topological category, we already know that the converse does not hold.

Reason: Let K be a knot which is topologically slice but smoothly non-slice. If K is ribbon, then K is smoothly slice. We have thus arrived at a contradiction.

In fact, Cochran, Orr, and Teichner [12] found many new topologically non-slice knots.

The Conway knot is topologically slice because the Alexander polynomial is 1. It was an open problem whether it is smooth slice before Piccirillo [112] proved very recently that the Conway knot is not slice.

8.2 Slice n-Dimensional Knots in $\mathbb{R}^{n+2}$

Generalizing Definition 8.1, we define *slice* n-knots in $\mathbb{R}^{n+2}$, where n is any positive integer. These are n-knots in $\mathbb{R}^{n+2}$ which bound $(n+1)$-discs in $\mathbb{R}^{n+3}_{\geqq 0}$. Although there are infinitely many non-slice 1-knots in $\mathbb{R}^3$, we have the following theorem.

Theorem 8.7 (Kervaire [51]). *All even-dimensional knots are slice.*

Levine characterized which $(2n+1)$-knots are slice and which are not, where n is a positive integer. See [58–61].

However, we have an outstanding open question in the case of high-dimensional link concordance. See the following section.

8.3 Slice n-Dimensional Links in $\mathbb{R}^{n+2}$

Generalizing Definition 8.1, we define *slice 1-links.* These are 1-links in $\mathbb{R}^3$ which bound disjoint discs in $\mathbb{R}^4_{\geqq 0}$.

The Hopf link is an example of a link which is not slice. In Figure 8.3, we see a non-trivial 2-component link which is slice. The proof that this link is in fact slice is left as an exercise for the reader. Hint: Try to draw a picture like Figure 8.1.

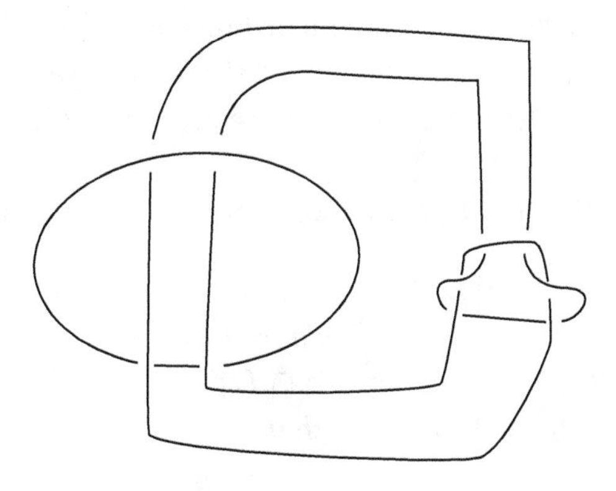

Figure 8.3 A slice link.

Generalizing slice 1-links, we define *slice n-links* in $\mathbb{R}^{n+2}$. Note that they bound disjoint $(n+1)$ discs in $\mathbb{R}^{n+3}_{\geqq 0}$.

As we cited in Theorem 8.7, all even-dimensional knots are slice. However, the analog for links is open:

Question 8.8. Are all even-dimensional links slice?

However, Cochran and Orr did prove the following exciting theorem about high-dimensional links.

Theorem 8.9 (Cochran and Orr [11]).

(1) *Let $m \geqq 1$. Not all $(2m+1)$-links are concordant to a boundary link.*
(2) *Not all 1-links with vanishing Milnor $\overline{\mu}$-invariants are concordant to a boundary link.*

L is a slice link if and only if L is concordant to the trivial link. For clarity, we provide a definition of concordance.

Let $[0,1]$ be the closed segment $\{0 \leqq t \leqq 1\}$. We call $S^n \times [0,1]$ an $(n+1)$-dimensional annulus. Take $\mathbb{R}^{n+2} \times [0,1]$. Let $L_i = (K_{i,1}, \ldots, K_{i,m})$ be an m-component n-link in $\mathbb{R}^{n+2}$ $(i = 1, 2)$. Take L_i in $\mathbb{R}^{n+2}$ at $t = i$ in $\mathbb{R}^{n+2} \times [0,1]$. Assume that there are m disjoint $(n+1)$-dimensional annuli C_i^{n+1} $(i = 1, \ldots, m)$ in $\mathbb{R}^{n+2} \times [0,1]$ such that ∂C_i^{n+1} is K_{i1} and K_{i2}. Then we say that L_1 and L_2 are *concordant*.

See [11] for more details on concordance and for definitions of the terminologies "boundary links" and "Milnor $\overline{\mu}$-invariants."

Cochran and Orr's result is fantastic, but Question 8.8 still remains.

The author showed a partial solution in Theorem 10.

Theorem 8.10 ([84]). *Let m be a positive integer. Let $L = (K_1, K_2)$ be a $2m$-link in $\mathbb{R}^{2m+2}$. There is a slice $(2m+1)$-disc D_i^{2m+1} for K_i $(i = 1, 2)$ such that $D_i^{2m+1} \cap D_j^{2m+1}$ in D_i^{2m+1} is the trivial one-component $(2m+1)$-link — that is, the trivial $(2m+1)$-knot.*

8.4 Projections of Knots

8.4.1 *An immersion which is the projection of no knot*

The *projection* of a 1-knot is the image of the knot under the map

$$\mathbb{R}^3 \to \mathbb{R}^2 : (x, y, z) \mapsto (x, y).$$

See Figure 8.4 for an example.

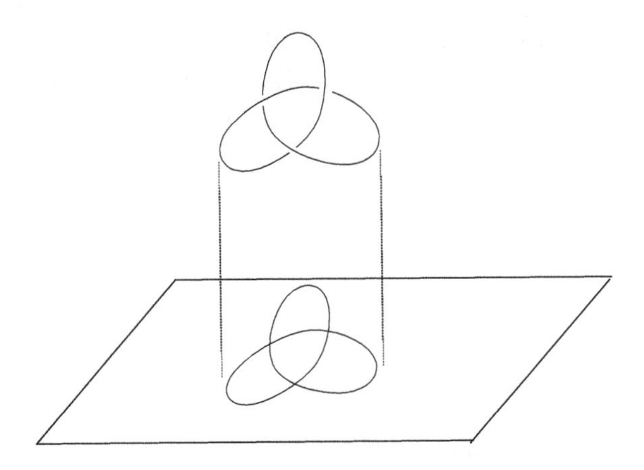

Figure 8.4 The projection of a one-dimensional knot.

Consider S^1 in $\mathbb{R}^2$, where S^1 may touch itself if it satisfies the condition: Self-intersections of S^1 are *transverse* — that is, self-intersections appear as shown in Figure 8.5.

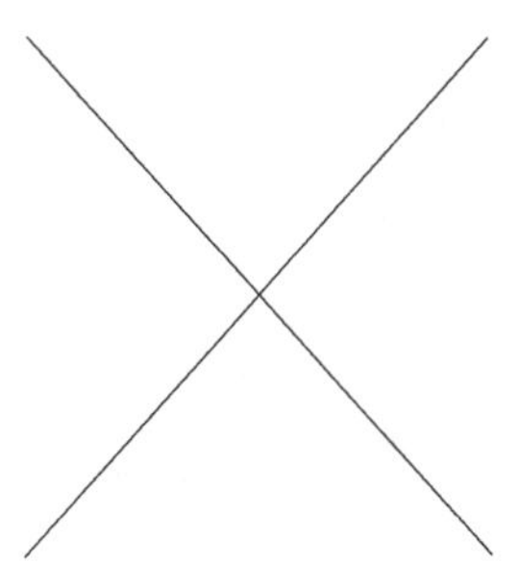

Figure 8.5 Two segments intersect transversely. Segments may be bent a little.

Furthermore, assume that the number of crossing points is finite. Under these conditions, we say that S^1 is *immersed transversely* into $\mathbb{R}^2$.

Note the difference among diagrams in Figure 1.2 of Chapter 1, projections and transverse immersions.

Question 8.11. Take any transverse immersion of a circle in $\mathbb{R}^2$. Is it the projection of a 1-knot in $\mathbb{R}^3$?

Answer: Yes. The proof is simple and is left as an exercise.

Now consider the sphere S^2 in $\mathbb{R}^3$, where we allow the sphere to touch itself under the condition that all self-intersections appear locally as what is shown in Figure 8.6.

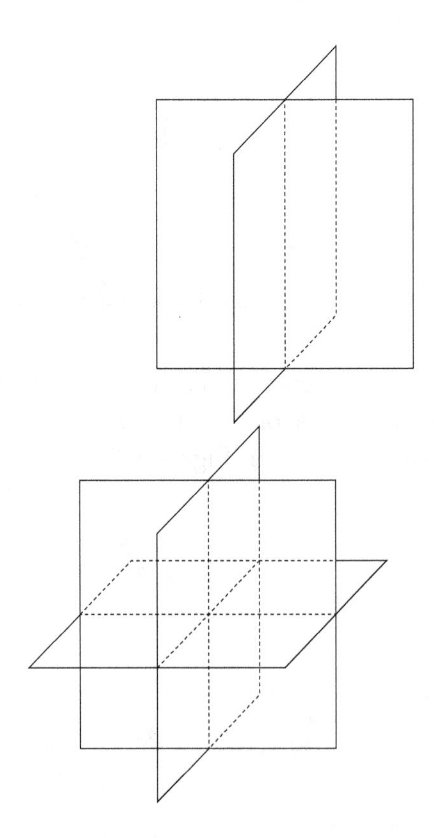

Figure 8.6 Two or three sheets intersect.

In Figure 8.6, two or three sheets intersect, where the sheets may be bent a little. We assume that the intersection is a union of a finite number of points, segments, and circles, where these segments may also be bent a little. Under these conditions, we say that S^2 is *immersed transversely* into $\mathbb{R}^3$. See textbooks which focus mainly on manifolds for a more in-depth exploration of transverse immersion.

We have defined the projection of a 1-knot, but we can define it for higher dimensions as well. The *projection* of a 2-knot K is the image of K under the map

$$\mathbb{R}^4 \to \mathbb{R}^3 : (x, y, z, w) \mapsto (x, y, z).$$

Question 8.12. Take an arbitrary transverse immersion of S^2 into $\mathbb{R}^3$. Is it the projection of a 2-knot?

Question 8.12 is a one-dimensional-higher analog of Question 8.11. Of course, there are countably infinitely many transverse immersions of S^2 into $\mathbb{R}^3$ which are the projections of 2-knots. However...
Answer: No.

Giller [27] proved it by using what we call *Boy's surface* which is drawn in Figure 8.7.

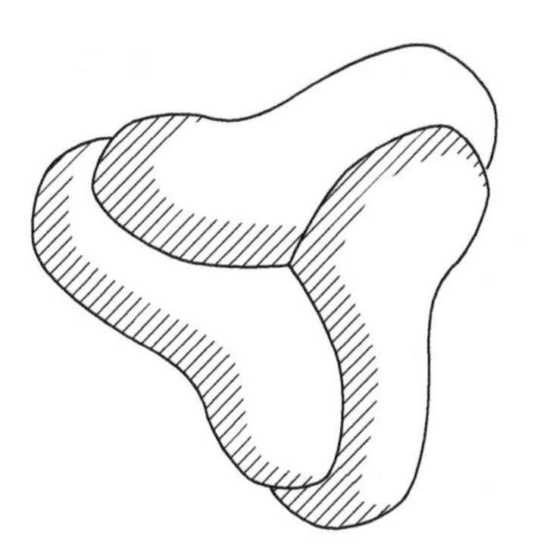

Figure 8.7 Boy's surface.

This interesting surface was discovered by Boy [6], the surface's namesake. In [74, page 121], Boy's paper is quoted.

Boy's surface, while just as important in topology as the Möbius band or the Klein bottle, is much less well-known. The author suspects that this is due to the more complicated shape of Boy's surface; despite being visualizable in $\mathbb{R}^3$, it is more difficult for beginners to imagine than the Möbius band and Klein bottle.

Thus, the author [94] has created a way to construct the Boy surface by using a pair of scissors, a piece of paper, and a strip of scotch tape. Readers may find a mini movie of this paper craft by typing "Eiji Ogasa" or "Make your Boy surface" into their search engines.

It is natural to ask a higher-dimensional analog of Question 8.12.

Question 8.13. Take a transverse immersion of the n-sphere S^n into $\mathbb{R}^{n+1}$. Is there an n-knot in $\mathbb{R}^{n+2}$ whose projection is the S^n in $\mathbb{R}^{n+1}$?

The *projection* of an n-knot K is the image of K by the map from $\mathbb{R}^{n+2}$ to $\mathbb{R}^{n+1}$; $(x_1, \ldots, x_{n+1}, x_{n+2}) \mapsto (x_1, \ldots, x_{n+1})$.

A *transverse immersion* S^n into $\mathbb{R}^{n+1}$ is a natural generalization of the lower dimensional one. See textbooks on manifolds for details.

The author [83] found that the answer to Question 8.13 is negative for all integers $n \geqq 3$. In proving this, he constructs an explicit transverse immersion of S^n into $\mathbb{R}^{n+1}$. This is a sub-theme of the author's paper [83]. The main result of [83] is introduced in the following section.

8.4.2 *An immersion which is the projection of a knot but not of an unknot*

Note the difference between the title of Section 8.4.1 and that of Section 8.4.2.

First, consider an easy question.

Question 8.14. Take a transverse immersion of S^1 into $\mathbb{R}^2$. It is the projection of a knot (see the answer to Question 8.11.). Is this the projection of a trivial 1-knot in $\mathbb{R}^3$?

Note the difference between Questions 8.11 and 8.14.

It is almost trivial that the answer is yes. Prove it!

We draw an example of this in Figure 8.8.

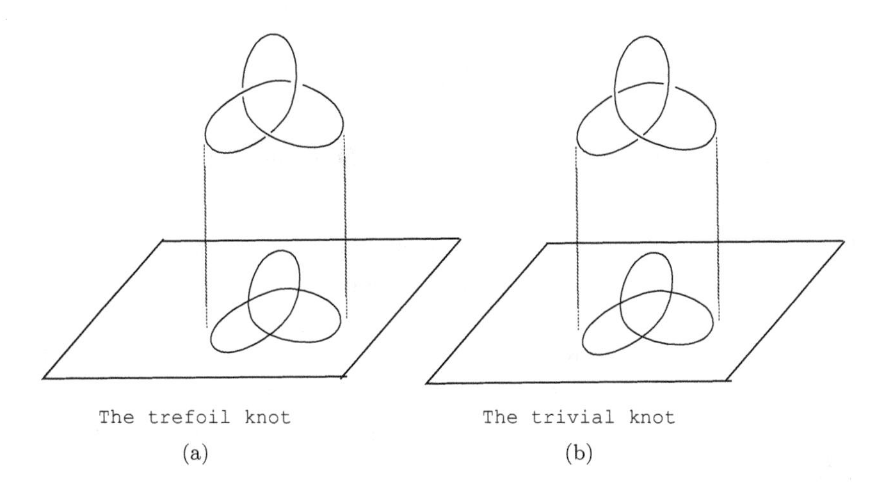

Figure 8.8 (a) A non-trivial 1-knot and (b) a trivial knot have the same projection.

It is very natural to ask a higher-dimensional analog of Question 8.14.

Question 8.15. Take an arbitrary transverse immersion P of the n-sphere S^n into $\mathbb{R}^{n+1}$. If P is a projection of an n-knot, is there a trivial n-knot whose projection is P?

Note the difference between Questions 8.13 and 8.15.

The $n = 1$ case is Question 8.14.

The author [82, 83] found that the answer to Question 8.15 is negative if $n \geqq 3$.

The author found an explicit way to make a transverse immersion of S^n into $\mathbb{R}^{n+1}$ which is the projection of a non-trivial n-knot but which is not the projection of any trivial n-knot. We use what we call a *K3 surface*, which you can find details on and images of in textbooks on algebraic geometry.

The $n = 2$ case of Question 8.15 is open. Try it out!

Another open question: The singularity of the transverse immersions in the author's papers [82, 83] consists of only double points. If $n \geqq 5$, the connected components of the singularity of the author's example in [83] are two. In the $n = 4, 3$ (or $= 2$), what is the least number of connected components of the singularity set when the set consists of only double points? Tackle this problem, too!

8.5 Can the Jones Polynomial Be Defined for Links in All Manifolds?

There is a natural one-to-one map from the set of links in $\mathbb{R}^3$ to that of links in S^3. Vaughan Jones [32] defined the Jones polynomial for links in S^3.

The Alexander polynomial and many other invariants for links in S^3 can easily be extended to 3-manifolds other than S^3. See Remark below Definition 7.2 of Chapter 7 for an overview of manifolds and notable examples.

It is natural to ask the following question.

Question 8.16. Can we extend the Jones polynomial for links in S^3 to the case of other 3-manifolds than S^3 in a consistent and meaningful way?

For now, we only have one partial answer, given by Kauffman in [41–43]. See Section 8.7.

Furthermore, we have the following open question: Can we define the Jones polynomial for n-dimensional knots in $\mathbb{R}^{n+2}$ (or other m-manifolds)? Here n and m are arbitrary positive integers.

8.6 Quantum Field Theory, Path Integrals, the Jones Polynomial, and Quantum Invariants of Closed Oriented 3-Manifolds

When Jones [32] introduced the Jones polynomial, he [32, page 360, §10] tried to define a 3-manifold invariant associated with the Jones polynomial, and succeeded in some cases.

After that, Witten [132] wrote a path integral to represent invariants of links in all (closed or compact) oriented 3-manifolds. If there is no link in a 3-manifold, the path integral represents a 3-manifold invariant. However, Witten's work was written from the perspective of theoretical physics, and was not completely mathematically rigorous.

Reshetikhin and Turaev [116] defined a 3-manifold invariant via surgery and quantum groups that one can view as a mathematically rigorous definition of the path integral. They defined it rigorously for all closed oriented 3-manifolds. Such 3-manifold invariants are called *quantum invariants.*

Although Witten wrote his famous path integral for links in all (closed or compact) oriented 3-manifolds, Question 8.16 is open. Witten only calculated the case of links in S^3 (see [132]), so the following question is open.

Question 8.17. Calculate Witten's path integral for links in any 3-manifold other than the 3-sphere.

Moreover, Witten [132] never answers Question 8.16.

What this means is that even at the physics level, the Jones polynomial has not been extended to all 3-manifolds. Witten only wrote a Lagrangian and an observable for a path integral in the case of links in all (closed or compact) oriented 3-manifolds. But that does not mean that the path integral has been calculated for links in all (closed or compact) oriented 3-manifolds.

In physics, we have the following: Even if we make a (seemingly) meaningful Lagrangian and an observable, the path integral associated with them cannot always be calculated. An example is the Witten path integral described above. Another unsolved problem is the following: It is not known to physics how to calculate the path integral if we replace the Chern–Simons 3-form on 3-manifolds with a Chern–Simons $(2p+1)$-form on $(2p+1)$-manifolds. In this case, p is any integer $\geqq 2$ in the Witten path integral without Wilson line for (closed) oriented $(2p+1)$ manifolds.

Indeed, nowadays physicists only calculate path integrals only when they can calculate them. If the path integral is not calculated explicitly, neither a mathematician nor a physicist regards the theory of the Lagrangian as a complete one.

Furthermore, even if we calculate path integrals, the result of the calculation is sometimes largely unexpected. See Lee and Yang's example [56] explained in [120, the last part of Section 5.1]: the

Feynman path integral does not induce the Schrödinger equation in some cases where it could be expected.

A current situation of QCD and a history of QED is as follows. Before Tomonaga, Feynman, and Schwinger discovered renormalization, we wrote a well-known Lagrangian and a path integral for QED. We have written a well-known Lagrangian and a path integral for QCD, but we cannot say matter-of-factly that we have completed the QCD case.

Reshetikhin and Turaev [116, Theorem 3.3.3, page 560] defined invariants for links in any closed oriented 3-manifold M by using their quantum invariants of closed oriented 3-manifolds. If $M = S^3$, these invariants are invariants of links in S^3. Apply the definition in [116, Theorem 3.3.3, page 560] strictly to the case of S^3. The reader may very easily come to understand that the definition of Reshetikhin–Turaev invariants of links in S^3 is different from that of the Jones polynomial of links in S^3. Reshetikhin–Turaev invariants depend on the use of the colored Jones polynomials at roots of unity and the use of Kirby calculus. See [55] for Kirby calculus.

From Reshetikhin and Turaev's theory, we get the following open question.

Question 8.18. Do Reshetikhin–Turaev invariants of links in S^3 recover the Jones polynomial of links in S^3?

Put otherwise, this question indicates that we are unsure whether the Reshetikhin–Turaev invariants of links in any 3-manifold are extensions of the Jones polynomial of links in S^3.

Neither Witten [132] nor Reshetikhin and Turaev [116, Theorem 3.3.3, page 560] answer Question 8.16.

Furthermore, in spite of Witten's great work and Reshetikhin and Turaev's exciting progress in this area, a quantum invariant for compact oriented 3-manifold with non-vacuous boundary had not even been calculated on the level of physics, let alone with mathematical rigor.

Following Witten and the Reshetikhin–Turaev team, Kauffman and the author [48] introduced such invariants. See Section 8.8.

8.7 Virtual Knots and Links

So far, we only have one partial answer to Question 8.16, and that is given by Kauffman. He extended the Jones polynomial to the case of (any oriented closed surface) $\times$ (the interval), by introducing *virtual links* [41–43], without using a path integral or another method of physics.

The set of virtual links is a quotient set of all links in all thickened closed oriented surfaces. See [41–43]. We refer to the links in $\mathbb{R}^3$ which we have seen throughout the text as *classical links*. Note that there is a natural one-to-one correspondence between the set of links in $\mathbb{R}^3$ and that of those in S^3 (respectively, the thickened sphere).

By the above reason, etc., virtual knot theory is important. See [31, 71].

8.8 Quantum Invariants of Oriented 3-Manifolds with Non-Vacuous Boundary

Although Witten wrote his famous path integral for compact oriented 3-manifolds with non-vacuous boundary and Reshetikhin and Turaev defined their quantum invarints for closed oriented 3-manifolds, as aforementioned, a quantum invariant for compact oriented 3-manifold with non-vacuous boundary has yet to be defined even on the level of physics.

After the work of Witten as well as Reshetikhin and Turaev discussed in Section 8.6, Kauffman and the author [48] introduced quantum invariants for compact oriented 3-manifold with non-vacuous boundary, where the boundary is a disjoint union of two identical surfaces. The invariants are the first, non-trivial, and calculable examples of quantum invariants of 3-manifolds with non-vacuous boundary.

They used Kauffman's theory of virtual links in order to circumvent the need to use a path integral or other methods from physics. They used the Dye–Kauffman invariant [15] of framed virtual links. Dye, Kauffman, and the author [17] calculated some examples of the above invariants introduced by Kauffman and the author.

Furthermore, by using the above invariants introduced by Kauffman and the author, we defined a new invariant of links in S^3. This means that we have discovered a new invariant of classical links using virtual knot theory.

8.9 Khovanov–Lipshitz–Sarkar Stable Homotopy Type and High-Dimensional Space

Let K be an n-knot, where n is any positive integer. The Alexander polynomial of K is characterized by using explicit figures as in Definition 7.2 of Chapter 7.

Some readers may ask: Can we characterize the Jones polynomial of 1-links in S^3 by using explicit figures?

Lipshitz and Sarkar [63] used enhanced Kauffman states (see [3, 38, 53] for enhanced Kauffman states), and provided such a characterization. There is a stable homotopy type of a CW complex $X = \bigvee_i X_i$, where $\bigvee_i X^i$ is the one point union of compact CW complexes X_i, for any given 1-link in S^3 such that the graded Euler number of X induces the coefficient of the Jones polynomial. Furthermore, the stable homotopy type is stronger than the Jones polynomial as a link invariant. See Seed's paper [122].

Any *CW complex* is a union of n-discs, where n runs over non-negative integers. Here, attaching maps of n-discs satisfies some conditions. See [74] for the definitions.

This leads to the following important open question.

Question 8.19. Can we define Khovanov–Lipshitz–Sarkar stable homotopy type for links in any other 3-manifold besides S^3?

As of now, we only have one consistent partial answer in the case of links in thickened surfaces. It is given by Kauffman, Nikonov, and the author [49, 50] — see [98, 99] for background on Khovanov–Lipshitz–Sarkar homotopy type. In those texts, knowledge of Khovanov homology [53] is important. As link invariants, the Jones polynomial is weaker than Khovanov homology, which is weaker than Khovanov–Lipshitz–Sarkar homotopy type.

8.10 Floer Homology

Heegaard Floer homology, knot Floer homology for classical links in closed oriented 3-manifolds, sutured Floer homology, and bordered Floer homology are all very strong invariants. By applying them, we can detect the unknot in closed oriented 3-manifolds and can investigate whether or not any given knot in S^3 is fibered. They are also useful when researching knot concordance.

Heegaard Floer homology is defined by Ozsváth and Szabó in [106, 107]. Knot Floer homology is defined by Ozsváth and Szabó, and by Rasmussen. See [68, 108, 110, 114].

Sutured Floer homology is defined by Juhász in [33]. See [25, 34] for application of sutured Floer homology. Sutured Floer homology includes a Heegaard Floer homology and knot Floer homology as special cases: knot Floer homology is defined for null-homologous knots in closed oriented 3-manifolds (see [108, the third line in Section 1] and [114, the second line of the paragraph right above Section 1.1]), whereas sutured Floer homology is defined for all knots in all compact oriented 3-manifolds.

Bordered Floer homology is defined by Lipshitz, Ozsváth, and Thurston in [66]. By assuming the appropriate conditions, sutured Floer homology and bordered Floer homology can each be retrieved from one another.

The Alexander polynomial of a classical knot K in S^3 is Reidemeister torsion, which is the graded Euler characteristic of knot Floer homology (respectively, sutured Floer homology) of K [33, 107]. It is much connected with Turaev τ [33, 107, 128, 129].

From Levine and Ruberman [57], we see a discussion of a relationship between 2-knots and Floer homology.

Juhász, Kauffman, and the author [35] made the first application of Floer homology to virtual links. In [35], they also introduced a new Alexander polynomial for virtual knots, which is called the *twin Alexander polynomial.*

8.11 The Poincaré Conjecture Is Still Open?

The Poincaré conjecture is very important. The n-dimensional topological (respectively, PL, smooth) Poincaré conjecture is as follows

($n \in \mathbb{N}$): If X is an n-dimensional simply connected homology sphere, is X homeomorphic (respectively, PL homeomorphic, diffeomorphic) to the standard n-sphere S^n? Many parts of the conjecture are solved by Poincaré, Milnor [73], Smale [123], Kervaire and Milnor [52], Newman [76], Sullivan [125], Freedman [21], Perelman [111], and Hill, Hopkins, and Ravenel [30]. See also great works of Casson, Novikov, Browder, Wall, Kirby, Siebenmann, Thurston, Stallings, Cappell, Shaneson, and Donaldson associated with the above exciting results.

However, some parts of the Poincaré conjecture are still open. Note that in the literature, "Poincaré conjecture" typically refers to the three-dimensional case, though you may see the term refer to the conjecture in all dimensions. We detail the unsolved parts of the conjecture as follows.

Poincaré conjecture

	Topological	PL	Smooth
$n = 1$	Solved affirmatively	Solved affirmatively	Solved affirmatively
$n = 2$	Solved affirmatively	Solved affirmatively	Solved affirmatively
$n = 3$	Solved affirmatively	Solved affirmatively	Solved affirmatively
$n = 4$	Solved affirmatively	Open	Open
$n \geqq 5$	Solved affirmatively	Solved affirmatively	Many parts are solved, but some parts are open In some cases, the answer is negative In other cases, the answer is positive

The work of Manolescu and Piccirillo [69] suggests that Rasmussen's s-invariant [115] made from Khovanov homology, and Lipshitz and Sarkar's refinement of Rasmussen's s-invariant [65] achieved by applying Khovanov–Lipshitz–Sarkar homotopy type may work to solve the four-dimensional smooth Poincaré conjecture. Both invariants are used to research knot concordance.

In other words, the four-dimensional smooth Poincaré conjecture is as follows: Is there an exotic smooth structure on S^4?

On the other hand, Donaldson *et al.* proved that there are infinitely many exotic smooth structures on *Dolgachev surfaces*. See [13, 14] and [55, III§3].

The following question is open and is related to the four-dimensional Poincaré conjecture.

Question 8.20. Is there a smooth 4-manifold which admits only a finite number of smooth structures?

8.12 Working in High Dimensions Is Necessary

Perhaps someone may think, "I will study knots in $\mathbb{R}^3$, so I do not need to learn about higher-dimensional spaces." However, we showed in Section 8.9 that we cannot avoid using high-dimensional figures. (Moreover, Khovanov–Lipshitz–Sarkar stable homotopy type is constructed from a union of high-dimensional discs, in general.)

Furthermore, high-dimensional figures are fascinating in their own right. All in all, high-dimensional mathematics is very important for all human beings.

If you solve any of the open questions in this textbook, you will be a mathematical legend. Go ahead, attack them!

References

In the following, we list many important resources, but there exist many other exciting papers on the topics we mention in this text. If the reader wants to discover new amazing facts, the list of papers we provide here is a great place to begin one's search.

[1] J. W. Alexander, Topological invariants of knots and links, *Trans. Am. Math. Soc.* 30(2) (1928) 275–306. doi:10.2307/1989123.

[2] M. M. Asaeda, J. H. Przytycki and A. S. Sikora, Categorification of the Kauffman bracket skein module of I-bundles over surfaces, *Algebr. Geom. Topol.* 4 (2004) 1177–1210.

[3] D. Bar-Natan, On Khovanov's categorification of the Jones polynomial, *Algebr. Geom. Topol.* 2(1) (2002) 337–370 (electronic). MR 1917056 (2003h:57014).

[4] J. A. Baldwin, On the spectral sequence from Khovanov homology to Heegaard Floer homology, *Int. Math. Res. Not.* 15 (2011) 3426–3470.

[5] J. A. Baldwin, A. S. Levine and S. Sarkar, Khovanov homology and knot Floer homology for pointed links, *J. Knot Theor. Ramif.* 26(02) (2017) 1740004.

[6] W. Boy, Über die Curvatura integra und die Topologie geschlossener Flächen, *Math. Ann.* 57 (1903) 151–184.

[7] S. E. Cappell and J. L. Shaneson, There exist inequivalent knots with the same complement, *Ann. Math.* 103 (1976) 349–353.

[8] T. D. Cochran and R. E. Gompf, Applications of Donaldson's theorems to classical knot concordance, homology 3-spheres and property P, *Topology* 27(4) (1988) 495–512.

[9] A. J. Casson and C. McA. Gordon, Cobordism of classical knots, *Prog. Math.* 62 (1975) 181–199.

[10] A. J. Casson and C. McA. Gordon, On slice knots in dimension three, *Proc. Symp. Pure Math.* 32 (1978) 39–53.

[11] T. D. Cochran and K. E. Orr, Not all links are concordant to boundary links, *Ann. Math.* 138 (1993) 519–554.

[12] T. D. Cochran, K. E. Orr and P. Teichner, Knot concordance, Whitney towers and L2-signatures, *Ann. Math.* 157 (2003) 433–519.

[13] S. Donaldson, An application of gauge theory to four-dimensional topology, *J. Differ. Geom.* 18 (1983) 279–315.

[14] S. Donaldson, Irrationality and the h-cobordism conjecture, *J. Differ. Geom.* 26(1) (1987) 141–168.

[15] H. A. Dye and L. H. Kauffman, Virtual knot diagrams and the Witten–Reshetikhin–Turaev invariant, *J. Knot Theor. Ramif.* 148(8) (2005) 1045–1075. arXiv:math/0407407.

[16] H. A. Dye A. Kaestner and L. H. Kauffman, Khovanov homology, Lee homology and a Rasmussen invariant for virtual knots, *J. Knot Theor. Ramif.* 26(3) (2017) 1741001-1–1741001-57.

[17] H. A. Dye, L. H. Kauffman and E. Ogasa, Quantum invariants of links and 3-manifolds with boundary defined via virtual links: Calculation of some examples. arXiv:2203.12797 [math.GT].

[18] M. Farber, Classification of simple knots. *Uspekhi Mat. Nauk*, 38(5) (1983), 59–106.

[19] M. Š. Farber, algebraic classification of some even-dimensional spherical knots. I *Trans. Am. Math. Soc.* 281 (1984) 507–527.

[20] M. Š. Farber, An algebraic classification of some even-dimensional spherical knots. II *Trans. Am. Math. Soc.* 281 (1984) 529–570.

[21] M. Freedman, The topology of four-dimensional manifolds, *J. Diff. Geom.* 17 (1982) 357–453.

[22] M. Freedman, The disk theorem for four-dimensional manifolds, *Proc. I.C.M. (Warsaw)* (1983) 647–663.

[23] M. Freedman and F. Quinn, *Topology of 4-Manifolds*, Princeton University Press, Princeton, NJ, 1990.

[24] P. Freyd, D. Yetter, J. Hoste, W. B. R. Lickorish, K. Millett and A. Ocneanu, A new polynomial invariant of knots and links, *Bull. Am. Math. Soc.* 12(2) (1985) 239–246. doi:10.1090/S0273-0979-1985-15361-3.

[25] S. Friedl, A. Juhász and J. Rasmussen, The decategorification of sutured Floer homology, *J. Topol.* 4(2) (2011) 431–478. arXiv:0903.5287.

[26] C. McA. Gordon, Knots in the 4-sphere, *Comm. Math. Helv.* 51 (1976) 585–596.

[27] C. Giller, Towards a classical knot theory for surfaces in $\mathbb{R}^4$, *Illinois J. Math.* 26(4) (1982) 591–631.

[28] P. M. Gilmer and C. Livingston, The Casson–Gordon invariant and link concordance, *Topology* 31 (1992) 475–492.

[29] A. Haefliger, Knotted $(4k-1)$-spheres in $6k$-space, *Ann. Math.* 75(3) (1962) 452–466.

[30] M. A. Hill, M. J. Hopkins and D. C. Ravenel, On the non-existence of elements of Kervaire invariant one, *Ann. Math.* 184 (2016) 1–262.

[31] D. P. Ilyutko and V. O. Manturov, Virtual Knots: The State of the Art, World Scientific Publishing Co. Pte. Ltd., 2012. English translation of [71].

[32] V. F. R. Jones, Hecke algebra representations of braid groups and link, *Ann. Math.* 126 (1987) 335–388.

[33] A. Juhász, Holomorphic discs and sutured manifolds, *Algebr. Geom. Topol.* 6(3) (2006) 1429–1457. arXiv:math/0601443.

[34] A. Juhász, Floer homology and surface decompositions, *Geom. Topol.* 12(1) (2008) 299–350. arXiv:math/0609779.

[35] A. Juhász, L. H. Kauffman and E. Ogasa, New invariants for virtual knots via spanning surfaces. arXiv:2207.08129 [math.GT].

[36] L. H. Kauffman and W. D. Neumann, Products of knots, branched fibrations and sums of singularities, *Topology* 16(4) (1977) 369–393.

[37] L. H. Kauffman, Products of knots, *Bull. Am. Math. Soc.* 80(6) (1974) 1104–1107.

[38] L. H. Kauffman, State models and the Jones polynomial, *Topology* 26 (1987) 395-407.

[39] L. H. Kauffman, On Knots, *Ann. Math. Stud.* 115 (1987).

[40] L. H. Kauffman, *Knots and Physics*, Second Edition, World Scientific Publishing, 1994.

[41] L. H. Kauffman, Talks at MSRI Meeting in January 1997, AMS Meeting at University of Maryland, College Park in March 1997, Isaac Newton Institute Lecture in November 1997, Knots in Hellas Meeting in Delphi, Greece in July 1998, APCTP-NANKAI Symposium on Yang-Baxter Systems, Non-Linear Models and Applications at Seoul, Korea in October 1998.

[42] L. H. Kauffman, Virtual knot theory, *Eur. J. Combinat.* 20 (1999), 663–691, Article No. eujc.1999.0314. Available online at http://www .idealibrary.commath/9811028[math.GT].

[43] L. H. Kauffman, Introduction to virtual knot theory, *J. Knot Theor. Ramif.* 21(13) (2012) 1240007, 37pp.

[44] L. H. Kauffman and E. Ogasa, Local moves on knots and products of knots, *Knots Poland III Banach Center Publ.* 103(1) (2014) 159–209. arXiv:1210.4667 [math.GT].

[45] L. H. Kauffman and E. Ogasa, Local moves on knots and products of knots II, *J. Knot Theor. Ramif.* 30(10) (2021) 2140006-1–2140006-44. arXiv:1406.5573 [math.GT].

[46] L. H. Kauffman and E. Ogasa, Brieskorn submanifolds, local moves on knots, and knot products, *J. Knot Theor. Ramif.* 28(10) (2019) 1950068-1–1950068-42. https://doi.org/10.1142/S0218216519500688, arXiv:1504.01229 [mathGT].

[47] L. H. Kauffman and E. Ogasa, Steenrod square for virtual links toward Khovanov–Lipshitz–Sarkar stable homotopy type for virtual links. arXiv:2001.07789 [math.GT].

[48] L. H. Kauffman and E. Ogasa, A new classical link invariant defined via virtual links and quantum invariants of 3-anifolds with boundary. arXiv:2108.13547 [math.GT].

[49] L. H. Kauffman, I. M. Nikonov and E. Ogasa, Khovanov–Lipshitz–Sarkar homotopy type for links in thickened higher genus surfaces, *J. Knot Theor. Ramif.* 30(8) (2021) 2150052-1–2150052-48. https//doi.org/10.1142/S0218216521500528, arXiv: 2007.09241 [math.GT].

[50] L. H. Kauffman, I. M. Nikonov and E. Ogasa, Khovanov–Lipshitz–Sarkar homotopy type for links in thickened surfaces and those in S^3 with new modulis. arXiv:2109.09245 [math.GT].

[51] M. Kervaire, Les noeudes de dimensions supéreures, *Bull. Soc. Math. France* 93 (1965) 225–271.

[52] M. A. Kervaire and J. W. Milnor, Groups of homotopy spheres: I, *Ann. Math.* 77(3) (1963) 504–537.

[53] M. Khovanov, A categorification of the Jones polynomial, *Duke Math. J.* 101(3) (2000) 359–426.

[54] S. Kinoshita, On the Alexander polynomials of 2-spheres in a 4-sphere, *Ann. Math.* 74 (1961) 518–531.

[55] R. C. Kirby, The topology of 4-manifolds, *Lecture Notes Math.* 1374 (1989).

[56] T. D. Lee and C. N. Yang, Theory of charged vector mesons interacting with the electromagnetic field, *Phys. Rev.* 128(2) (1962) 885.

[57] A. S. Levine and D. Ruberman, Heegaard Floer invariants in codimension one, *Trans. Am. Math. Soc.* 371 (2019) 3049–3081.

[58] J. Levine, Knot cobordism in codimension two, *Comment. Math. Helv.* 44 (1969) 229–244.

[59] J. Levine, An algebraic classification of some knots of codimension two, *Comment. Math. Helv.* 45 (1970) 185–198.

[60] J. Levine, Polynomial invariants of knots of codimension two, *Ann. Math.* 84 (1966) 537–554.

[61] J. Levine and K. Orr, A survey of applications of surgery to knot and link theory. *Surveys on Surgery Theory: Surveys Presented in Honor of C.T.C. Wall*, Vol. 1, pp. 345-364, Princeton University Press, Princeton, NJ, *Ann. of Math. Stud.* 145 (2000).

[62] J. Levine, Link invariants via the eta-invariant *Comment. Math. Helv.* 69 (1994) 82–119.

[63] R. Lipshitz and S. Sarkar, A Khovanov stable homotopy type, *J. Am. Math. Soc.* 27(4) (2014) 983–1042.

[64] R. Lipshitz and S. Sarkar, A Steenrod square on Khovanov homology, *J. Topol.* 7(3) (2014) 817–848.
[65] R. Lipshitz and S. Sarkar, A refinement of Rasmussen's s-invariant, *Duke Math. J.* 163(5) (2014) 923–952.
[66] R. Lipshitz, P. Ozsváth and D. Thurston, Bordered Heegaard Floer homology: Invariance and pairing, 2008, Preprint, arXiv:0810.0687, *Mem. Am. Math. Soc.* 254 (2018) 1216.
[67] C. Manolescu and P. S. Ozsvath, On the Khovanov and knot Floer homologies of quasi-alternating links. arXiv:0708.3249 [math.GT].
[68] C. Manolescu, P. S. Ozsváth, Z. Szabó and D. Thurston, On combinatorial link Floer homology, *Geom. Topol.* 11 (2007) 2339–2412.
[69] C. Manolescu and L. Piccirillo, From zero surgeries to candidates for exotic definite four-manifolds. arXiv:2102.04391.
[70] V. O. Manturov, Khovanov homology for virtual links with arbitrary coeficients, *J. Knot Theor. Ramif.* 16(3) (2007) 345–377.
[71] V. O. Manturov, Virtual knots: The state of the art, (2010) (in Russian).
[72] V. O. Manturov and I. M. Nikonov, Homotopical Khovanov homology, *J. Knot Theor. Ramif.* 24 (2015) 1541003.
[73] J. W. Milnor, On manifolds homeomorphic to the 7-sphere, *Ann. Math.* 64 (1956) 399–405.
[74] J. W. Milnor and J. D. Stasheff, Characteristic classes, *Ann. Math. Stud.* 76 (1974).
[75] J. R. Munkres, *Elements of Algebraic Topology*, Westview Press, 1996.
[76] M. H. A. Newman, "The engulfing theorem for topological manifolds" *Ann. Math.* 84 (1966) 555–571.
[77] I. M. Nikonov, Virtual index cocycles and invariants of virtual links. arXiv:2011.00248.
[78] E. Ogasa, The intersection of spheres in a sphere and a new geometric meaning of the Arf invariants, *J. Knot Theory Ramif.* 11 (2002) 1211–1231. Univ. of Tokyo preprint series UTMS 95-7, math.GT/0003089, http://xxx.lanl.gov.
[79] E. Ogasa, Intersectional pairs of n-knots, local moves of n-knots and invariants of n-knots, *Math. Res. Lett.* 5 (1998) 577–582. Univ. of Tokyo preprint series UTMS 95-50.
[80] E. Ogasa, The intersection of spheres in a sphere and a new application of the Sato-Levine invariant, *Proc. Am. Math. Soc.* 126 (1998) 3109–3116. UTMS 95-54.
[81] E. Ogasa, Some properties of ordinary sense slice 1-links: Some answers to the problem (26) of Fox, *Proc. Am. Math. Soc.* 126 (1998) 2175–2182. UTMS 96-11.

[82] E. Ogasa, The projections of n-knots which are not the projection of any unknotted knot, *J. Knot Theor. Ramif.* 10(1) (2001) 121–132. UTMS 97-34, math.GT/0003088.

[83] E. Ogasa, Singularities of projections of n-dimensional knots, *Math. Proc. Camb. Philos. Soc.* 126 (1999) 511–519, UTMS 96-39.

[84] E. Ogasa, Link cobordism and the intersection of slice discs, *Bull. Lond. Math. Soc.* 31 (1999) 1–8.

[85] E. Ogasa, Ribbon-moves of 2-links preserve the μ-invariant of 2-links, *J. Knot Theor Ramif.* 13(5) (2004) 669–687. UTMS 97-35, math.GT/0004008.

[86] E. Ogasa, Supersymmetry, homology with twisted coefficients and n-dimensional knots, *Int. J. Mod. Phys. A* 21(19–20) (2006) 4185–4196. hep-th/0311136.

[87] E. Ogasa, Local move identities for the Alexander polynomials of high-dimensional knots and inertia groups, *J. Knot Theor. Ramif.* 18(4) (2009) 531–545, UTMS 97-63 math.GT/0512168.

[88] E. Ogasa, Nonribbon 2-links all of whose components are trivial knots and some of whose bund-sums are nonribbon knots, *J. Knot Theor. Ramif.* 10(6) (2001) 913–922.

[89] E. Ogasa, n-dimensional links, their components, and their band-sums. UTMS00-65, math.GT/0011163.

[90] E. Ogasa, Ribbon-moves of 2-knots: The Farber–Levine pairing and the Atiyah–Pathodi–Singer–Casson–Gordon–Ruberman $\tilde{\eta}$ invariant of 2-knots, *J. Knot Theor. Ramif.* 16(5) (2007) 523–543. math.GT/0004007, UTMS 00-22, math.GT/0407164.

[91] E. Ogasa, Supersymmetry, homology with twisted coefficients and n-dimensional knots, *Int. J. Mod. Phys. A* 21(19–20) (2006) 4185–4196. hep-th/0311136.

[92] E. Ogasa, A new invariant associated with decompositions of manifolds. math.GT/0512320, hep-th/0401217.

[93] E. Ogasa, A new obstruction for ribbon-moves of 2-knots: 2-knots fibred by the punctured 3-torus and 2-knots bounded by the Poincaré sphere. arXiv:1003.2473math.GT.

[94] E. Ogasa, Make your Boy surface. arXiv:1303.6448 math.GT.

[95] E. Ogasa, Local-move-identities for the $Z[t, t^{-1}]$-Alexander polynomials of 2-links, the alinking number, and high dimensional analogues. arXiv:1602.07775.

[96] E. Ogasa, A new pair of non-cobordant surface-links which the Orr invariant, the Cochran sequence, the Sato–Levine invariant, and the alinking number cannot find. arXiv:1605.06874.

[97] E. Ogasa, Ribbon-move-unknotting-number-two 2-knots, pass-move-unknotting-number-two 1-knots, and high dimensional analogue. arXiv:1612.03325.

[98] E. Ogasa, An elementary introduction to Khovanov–Lipshitz–Sarkar stable homotopy type. The readers can find this article by typing in the title in search engine.

[99] E. Ogasa, Easy examples of explicit construction of Khovanov–Lipshitz–Sarkar stable homotopy type,

[100] E. Ogasa, *Yojigen ijou no kûkan ga mieru* (in Japanese), Beret Shuppan, 2006.

[101] E. Ogasa, *Ijigen e no tobira* (in Japanese), Nippon Hyoron Sha Co., Ltd., 2009.

[102] E. Ogasa, *Sotaiseiriron no shiki wo michibiite miyo, sosite hito ni hanaso* (in Japanese), Beret Shuppan, 2013.

[103] E. Ogasa, *Kojigen kukan wo miru hoho* (in Japanese), Kodansha Ltd., 2019.

[104] E. Ogasa, *Tayotai towa nanika* (in Japanese), Kodansha Ltd., 2021.

[105] K. E. Orr, New link invariants and applications *Comment. Math. Helv.* 62 (1987) 542–560.

[106] P. S. Ozsváth and Z. Szabó, Holomorphic Disks and Topological Invariants for Closed Three-Manifolds, *Annals of Mathematics Second Series* 159 (2004) 1027–1158.

[107] P. S. Ozsváth and Z. Szabó, Holomorphic disks and three-manifold invariants: Properties and applications, *Ann. Math.* 159(3) (2004) 1159–1245.

[108] P. S. Ozsváth and Z. Szabó, Holomorphic disks and knot invariants, *Adv. Math.* 186(1) (2004) 58–116.

[109] P. S. Ozsváth and Z. Szabó, Holomorphic disks and genus bounds, *Geom. Topol.* 8 (2004) 311–334.

[110] P. S. Ozsváth, A. I. Stipsicz and Z. Szabó, *Grid homology for knots and links, (Mathematical Surveys and Monographs)*, American Matematical Society 2015.

[111] G. Perelman, "The entropy formula for the Ricci flow and its geometric applications (November 11, 2002), arXiv:math.DG/0211159. Ricci flow with surgery on three-manifolds (March 10, 2003), arXiv:math.DG/0303109. Finite extinction time for the solutions to the Ricci flow on certain three-manifolds (July 17, 2003), arXiv: math.DG/0307245.

[112] L. Piccirillo, The Conway knot is not slice. arXiv:1808.02923.

[113] J. H. Przytycki and P. Traczyk, Invariants of links of Conway type, *Kobe J. Math.* 4 (1987) 115–139.

[114] J. Rasmussen, Floer homology and knot complements, PhD thesis, Harvard University, 2003.

[115] J. Rasmussen, Khovanov homology and the slice genus, *Invent. Math.* 182 (2010) 419–447.

[116] N. Reshetikhin and V. G. Turaev, Invariants of 3-manifolds via link polynomials and quantum groups, *Invent. Math.* 103 (1991) 547–597.
[117] D. Rolfsen, *Knots and Links*, Publish or Perish, Inc., 1976.
[118] D. Ruberman, Doubly slice knots and the Casson–Gordon invariants, *Trans. Am. Math. Soc.* 279 (1983) 569–588.
[119] W. Rushworth, Doubled Khovanov homology, *Can. J. Math.* 70 (2018) 1130–1172.
[120] L. H. Ryder, *Quantum Field Theory*, Second Edition, Cambridge University Press, 1996.
[121] N. Sato, Cobordisms of semi-boundary links, *Topol. Appl.* 18 (1984) 225–234.
[122] C. Seed, Computations of the Lipshitz–Sarkar Steenrod square on Khovanov homology. arXiv:1210.1882.
[123] S. Smale, Generalized Poincaré's conjecture in dimensions greater than four, *Ann. Math.* 74 (1961) 391–406.
[124] A. I. Suciu, Inequivalent frame-spun knots with the same complement, *Comment. Math. Helv.* 67 (1992) 47–63.
[125] D. Sullivan, On the Hauptvermutung for manifolds, *Bull. Am. Math. Soc.* 73(4) (1967) 598–600.
[126] H. F. Trotter, Non-invertible knots exist, *Topology* 2 (1963) 275–280.
[127] D. Tubbenhauer, Virtual Khovanov homology using cobordisms, *J. Knot Theor. Ramif.* 23 (2014), 1450046, 91pp.
[128] V. G. Turaev, *Introduction to Combinatorial Torsions*, Notes taken by Felix Schlenk, Basel; Birkhäuser, Basel, Boston, Berlin, 2001, Springer, Lectures in Mathematics.
[129] V. G. Turaev, Reidemeister torsion and the Alexander polynomial, *Mat. USSSR Sb.* 101 (1976) 252–270.
[130] Viro, Khovanov homology of Signed diagrams, 2006 (an unpublished note).
[131] E. Zeeman, Twisting spun knots, *Trans. AMS* 115 (1965) 471–495.
[132] E. Witten, Quantum field theory and the Jones polynomial, *Comm. Math. Phys.* 121 (1989) 351–399.

Index

Vol. 68: *Mathematics of Harmony as a New Interdisciplinary Direction and "Golden" Paradigm of Modern Science*
Volume 2: Algorithmic Measurement Theory, Fibonacci and Golden Arithmetic's and Ternary Mirror-Symmetrical Arithmetic
by A. Stakhov

Vol. 67: On Complementarity: A Universal Organizing Principle
by J. S. Avrin

Vol. 66: *Invariants and Pictures: Low-dimensional Topology and Combinatorial Group Theory*
by V. O. Manturov, D. Fedoseev, S. Kim & I. Nikonov

Vol. 65: *Mathematics of Harmony as a New Interdisciplinary Direction and "Golden" Paradigm of Modern Science*
Volume 1: The Golden Section, Fibonacci Numbers, Pascal Triangle, and Platonic Solids
by A. Stakhov

Vol. 64: *Polynomial One-cocycles for Knots and Closed Braids*
by T. Fiedler

Vol. 63: *Board Games: Throughout the History and Multidimensional Spaces*
by J. Kyppö

Vol. 62: *The Mereon Matrix: Everything Connected through (K)nothing*
edited by J. B. McNair, L. Dennis & L. H. Kauffman

Vol. 61: *Numeral Systems with Irrational Bases for Mission-Critical Applications*
by A. Stakhov

Vol. 60: *New Horizons for Second-Order Cybernetics*
edited by A. Riegler, K. H. Müller & S. A. Umpleby

Vol. 59: *Geometry, Language and Strategy: The Dynamics of Decision Processes — Vol. 2*
by G. H. Thomas

Vol. 58: *Representing 3-Manifolds by Filling Dehn Surfaces*
by R. Vigara & A. Lozano-Rojo

Vol. 57: *ADEX Theory: How the ADE Coxeter Graphs Unify Mathematics and Physics*
by S.-P. Sirag

Vol. 56: *New Ideas in Low Dimensional Topology*
edited by L. H. Kauffman & V. O. Manturov

Vol. 55: *Knots, Braids and Möbius Strips*
by J. Avrin

Vol. 54: *Scientific Essays in Honor of H Pierre Noyes on the Occasion of His 90th Birthday*
edited by J. C. Amson & L. H. Kauffman

Vol. 53: *Knots and Physics (Fourth Edition)*
by L. H. Kauffman

Vol. 52: *Algebraic Invariants of Links (Second Edition)*
by J. Hillman

Vol. 51: *Virtual Knots: The State of the Art*
by V. O. Manturov & D. P. Ilyutko

Vol. 50: *Topological Library*
Part 3: Spectral Sequences in Topology
edited by S. P. Novikov & I. A. Taimanov

Vol. 49: *Hopf Algebras*
by D. E. Radford

Vol. 48: *An Excursion in Diagrammatic Algebra: Turning a Sphere from Red to Blue*
by J. S. Carter

Vol. 47: *Introduction to the Anisotropic Geometrodynamics*
by S. Siparov

Vol. 46: *Introductory Lectures on Knot Theory*
edited by L. H. Kauffman, S. Lambropoulou, S. Jablan & J. H. Przytycki

Vol. 45: *Orbiting the Moons of Pluto*
Complex Solutions to the Einstein, Maxwell, Schrödinger and Dirac Equations
by E. A. Rauscher & R. L. Amoroso

Vol. 44: *Topological Library*
Part 2: Characteristic Classes and Smooth Structures on Manifolds
edited by S. P. Novikov & I. A. Taimanov

Vol. 43: *The Holographic Anthropic Multiverse*
by R. L. Amoroso & E. A. Ranscher

Vol. 42: *The Origin of Discrete Particles*
by T. Bastin & C. Kilmister

Vol. 41: *Zero to Infinity: The Fountations of Physics*
by P. Rowlands

Vol. 40: *Intelligence of Low Dimensional Topology 2006*
edited by J. Scott Carter et al.

Vol. 39: *Topological Library*
Part 1: Cobordisms and Their Applications
edited by S. P. Novikov & I. A. Taimanov

Vol. 38: *Current Developments in Mathematical Biology*
edited by K. Mahdavi, R. Culshaw & J. Boucher

Vol. 37: *Geometry, Language, and Strategy*
by G. H. Thomas

Vol. 36: *Physical and Numerical Models in Knot Theory*
edited by J. A. Calvo et al.

Vol. 35: *BIOS — A Study of Creation*
by H. Sabelli

Vol. 34: *Woods Hole Mathematics — Perspectives in Mathematics and Physics*
edited by N. Tongring & R. C. Penner

Vol. 33: *Energy of Knots and Conformal Geometry*
by J. O'Hara

Vol. 32: *Algebraic Invariants of Links*
by J. A. Hillman

Vol. 31: *Mindsteps to the Cosmos*
by G. S. Hawkins

Vol. 30: *Symmetry, Ornament and Modularity*
by S. V. Jablan

Vol. 29: *Quantum Invariants — A Study of Knots, 3-Manifolds, and Their Sets*
by T. Ohtsuki

Vol. 28: *Beyond Measure: A Guided Tour Through Nature, Myth, and Number*
by J. Kappraff

Vol. 27: *Bit-String Physics: A Finite and Discrete Approach to Natural Philosophy*
by H. Pierre Noyes; edited by J. C. van den Berg

Vol. 26: *Functorial Knot Theory — Categories of Tangles, Coherence, Categorical Deformations, and Topological Invariants*
by David N. Yetter

Vol. 25: *Connections — The Geometric Bridge between Art and Science (Second Edition)*
by J. Kappraff

Vol. 24: *Knots in HELLAS '98 — Proceedings of the International Conference on Knot Theory and Its Ramifications*
edited by C. McA Gordon, V. F. R. Jones, L. Kauffman, S. Lambropoulou & J. H. Przytycki

Vol. 23: *Diamond: A Paradox Logic (Second Edition)*
by N. S. Hellerstein

Vol. 22: *The Mathematics of Harmony — From Euclid to Contemporary Mathematics and Computer Science*
by A. Stakhov (assisted by S. Olsen)

Vol. 21: *LINKNOT: Knot Theory by Computer*
by S. Jablan & R. Sazdanovic

Vol. 20: *The Mystery of Knots — Computer Programming for Knot Tabulation*
by C. N. Aneziris

Vol. 19: *Ideal Knots*
by A. Stasiak, V. Katritch & L. H. Kauffman

Vol. 18: *The Self-Evolving Cosmos: A Phenomenological Approach to Nature's Unity-in-Diversity*
by S. M. Rosen

Vol. 17: *Hypercomplex Iterations — Distance Estimation and Higher Dimensional Fractals*
by Y. Dang, L. H. Kauffman & D. Sandin

Vol. 16: *Delta — A Paradox Logic*
by N. S. Hellerstein

Vol. 15: *Lectures at KNOTS '96*
by S. Suzuki

Vol. 14: *Diamond — A Paradox Logic*
by N. S. Hellerstein

Vol. 13: *Entropic Spacetime Theory*
by J. Armel

Vol. 12: *Relativistic Reality: A Modern View*
edited by J. D. Edmonds, Jr.

Vol. 11: *History and Science of Knots*
edited by J. C. Turner & P. van de Griend

Vol. 10: *Nonstandard Logics and Nonstandard Metrics in Physics*
by W. M. Honig

Vol. 9: *Combinatorial Physics*
by T. Bastin & C. W. Kilmister

Vol. 8: *Symmetric Bends: How to Join Two Lengths of Cord*
by R. E. Miles

Vol. 7: *Random Knotting and Linking*
edited by K. C. Millett & D. W. Sumners

Vol. 6: *Knots and Applications*
edited by L. H. Kauffman

Vol. 5: *Gems, Computers and Attractors for 3-Manifolds*
by S. Lins

Vol. 4: *Gauge Fields, Knots and Gravity*
by J. Baez & J. P. Muniain

Vol. 3: *Quantum Topology*
edited by L. H. Kauffman & R. A. Baadhio

Vol. 2: *How Surfaces Intersect in Space — An Introduction to Topology (Second Edition)*
by J. S. Carter

Vol. 1: *Knots and Physics (Third Edition)*
by L. H. Kauffman

CPSIA information can be obtained
at www.ICGtesting.com
Printed in the USA
BVHW051955260723
667859BV00002B/46